Charles Redway Wilmarth Dryer

Studies in Indiana Geography

Charles Redway Wilmarth Dryer

Studies in Indiana Geography

ISBN/EAN: 9783742809322

Manufactured in Europe, USA, Canada, Australia, Japa

Cover: Foto ©Klaus-Uwe Gerhardt /pixelio.de

Manufactured and distributed by brebook publishing software (www.brebook.com)

Charles Redway Wilmarth Dryer

Studies in Indiana Geography

STUDIES

IN

INDIANA GEOGRAPHY

———

EDITED BY

CHARLES REDWAY DRYER, M. A., M. D.

Professor of Geography in The Indiana State Normal School

FIRST SERIES

Terre Haute, Ind.
THE INLAND PUBLISHING COMPANY
1897

Copyright, 1897
BY
THE INLAND PUBLISHING COMPANY

TO
WILLIAM MORRIS DAVIS
PROFESSOR OF PHYSICAL GEOGRAPHY IN HARVARD UNIVERSITY,
TO WHOSE SUGGESTION THIS AND SO MANY OTHER EFFORTS FOR
THE IMPROVEMENT OF GEOGRAPHIC TEACHING ARE DUE,
THESE STUDIES ARE RESPECTFULLY DEDICATED BY
THE EDITOR

TABLE OF CONTENTS

I. To Students and Teachers. 7

II. The New Geography. 9
          CHARLES R. DRYER.

III. The General Geography of Indiana. 17
          CHARLES R. DRYER.

IV. The Glacial Deposits of Indiana. 29
          FRANK LEVERETT.

V. The Erie-Wabash Region. 43
          CHARLES R. DRYER.

VI. The Morainic Lakes of Indiana. 53
          CHARLES R. DRYER.

VII. The Natural Resources of Indiana. 61
          WILLIS S. BLATCHLEY.

VIII. Indiana: A Century of Changes in the Aspects of Nature. . . 72
          AMOS W. BUTLER.

IX. A Study of the City of Terre Haute. 82
          CHARLES R. DRYER.

X. A Short History of the Great Lakes. 90
          FRANK BURSLEY TAYLOR.

LIST OF MAPS

Topographic Map of Indiana . . . . Frontispiece
Rainfall and Mean Temperature for 1896 25
Resources and Population 26
Glacial Map of Indiana 28
Glacial Map of North America 33
The Erie-Wabash Region 42
Morainic Lakes of Indiana . . . . Facing page 53
Two Stages of the Earlier Glacial Lakes . . . . 98 and 99
Glacial Lakes Algonquin and Iroquois 103
The Nipissing Great Lakes and the Champlain Sea 107

I.—TO STUDENTS AND TEACHERS OF GEOGRAPHY

To the student of scientific geography Indiana has been almost a sealed book. Very little of its area has ever been studied in the light of modern geographic science. The knowledge incidentally acquired by naturalists and geologists in the prosecution of their work is so scattered and buried in a mass of other material as to be unavailable. The paragraphs or chapters upon Indiana in the current text-books of geography are, for the most part, meagre, empty and uninteresting. One of the very latest and best primary geographies tells the children that "Indiana is hilly in the southeast, the rest of the state is one vast rolling prairie." There is as much geography to the square mile in Indiana as in any other state, and to Indiana students and teachers it is far more important than Thibet or Central Africa.

In an address before the National Geographic Society at Washington, February 3, 1893, upon the Improvement of Geographical Teaching, Professor William Morris Davis of Harvard University said: "The improvements needed in teaching geography in our schools involve a fuller investigation of the facts of the subject, a better knowledge of these facts by teachers, and a more skillful use of them in the process of teaching. I may briefly state my belief that skillful teaching goes along closely with fulness of knowledge. The third need will therefore be largely cared for when the second is supplied; but fulness of knowledge cannot be expected of a teacher while her understanding of the geographical features of the world and of our own country, and of the home state in particular, is gained only from the impoverished statements of the ordinary text-books, and while the original sources in which she may seek additional information are generally so few, so inaccessible, and so far below the standards of modern geographical research.

"It may not be generally recognized that there is still great need of exploration at home. It is not only in the farther corners of the world that discoveries are to be made. Nearly every state in our country must be much more carefully studied than it yet has been before its physical features will be made known to us. The geographical descriptions now accessible in print would be very gently characterized if only called 'old fashioned.' Where newer material has been published, it is generally fragmentary, brief, and imperfectly illustrated. The first elements of geographical study, the physical features of the earth—especially of its surface—still call for devoted investigation.

"Teachers in our public schools are well aware that they have not now the fuller account of the facts that they would enjoy; and yet they know not where to turn to find what they need. Many teachers, principals and superintendents admit at once that the books to which they now have access are quite insufficient to satisfy their wants, and they listen gladly to any feasible plan that will provide a more extended description and explanation of the facts of geography near at home, with which they have to deal from their earliest to their latest teaching. Geologists or geographers who are already acquainted with our local geography from personal experience can perform a grateful service to the schools by preparing elementary accounts of the regions with which they are familiar, and such books as these should be greatly multiplied; but, so far as I have been able to learn, it is only the smaller part of our country that is now known well enough to those who can be prevailed upon to write elementary books, and hence the importance of actual geographical exploration in order to supply our teachers with what they need. If some such plan as the one proposed above were put in operation, it might come to pass in a decade or two that the graduates of our common schools would not be so blinded as they now are to the facts of their home geography."*

To act upon the suggestion of Professor Davis, and to make a beginning toward a better state of things, this series of Studies in Indiana Geography has been planned and executed. The welcome accorded to them as they appeared in THE INLAND EDUCATOR has encouraged the editor and publisher to put them in a more convenient form for the school and the library. It is hoped that they may stimulate not only better teaching but further investigation. There is scarcely a township in Indiana which does not offer a field for new discovery, or at least for renewed and improved description. The Outline for Township Institute work for 1896-97 called for an essay on the Wabash river, and several thousand teachers discovered that there is less available knowledge about the Wabash than about the Congo, the LaPlata or the Hoang Ho. Geographic problems worthy of attention lie at everyone's door, and he who will undertake their study and report his results will strengthen himself and make a valuable contribution to geographic and educational science. It is hoped that good material may accumulate for a continuation of this series in the future. The thanks of the editor and the public are due to Messrs. Leverett, Blatchley, Butler, and Taylor for their generous and disinterested contributions to the cause of sound knowledge and good teaching. The article of Mr. Taylor has been rewritten to embody the latest results of investigation, and his maps have been redrawn.

*National Geographic Magazine, Vol. IV, p. 68.

II.—THE NEW GEOGRAPHY*

CHARLES R. DRYER

The prevalent and official view of geography held in Indiana to-day, is that it treats of the earth viewed in relation to the institutional life of man. I have been unable to discover the author of this particular formula, but the root of the idea undoubtedly lies in the *Erdkunde* of Karl Ritter, who taught geography at Berlin from 1820 to 1859. While Ritter did great service to geography by counteracting the tendency of Humboldt to include in it the whole of natural science, it is also true that his influence led to the other extreme, equally vicious, because it narrowed the science to a single relation and reduced it to little more than an adjunct to history. In a large number of American schools to-day, geography is still classified as "historical science," and a divorce between "geography" and "physical geography" is maintained, which in itself is absurd and destructive of the higher values of the science. To characterize this view as "old fashioned," antiquated and out of date, is to use very mild language. It looks like a willful shutting of eyes to the scientific progress of the last half-century, of which geography has enjoyed its share. It sounds like a survival from the old geocentric theory of the universe, prolonging into the twentieth century the idea that the universe was made for man and has no meaning apart from him. Its implication is strongly teleological, and the problem which Ritter and Guyot attacked with such brilliant ability, was nothing less than to show how each continent has been especially designed to produce and foster a special type of human culture. The Ritterian idea is at bottom a religious rather than a scientific one; and in the light of modern evolutionary thought such problems assume a different aspect. They are seen to be, for the present at least, beyond human knowledge and ability, and the modern student prefers to devote himself to the purely physical problems. The center of gravity is shifting away from man to nature. Physical geography, enlarged and enriched by whole new sciences, is being restored to its proper place at the foundation, and modern geography is becoming more a natural than a historical science. A few recent expressions of opinion from eminent and, so far as may be, authoritative German, French and English sources, will serve to present the modern views.

* Read before the Indiana State Teachers' Association, December, 1896.

The Third International Geographical Congress, which met at Venice in 1881, adopted the following resolutions:

(A) "The scientific object of geography comprehends the study of the superficial forms of the earth; it extends also to the reciprocal relations of the different branches of the organic world.

(B) "That which eminently distinguishes geography from the auxiliary sciences is that it localizes objects; that is to say, it indicates in a positive and constant manner the distribution of beings, organic and inorganic, upon the earth."

Professor Hettner of the University of Leipsic, wrote in 1895: "The geography of to-day starts from the point of view of diversity in space, and aims at a scientific explanation of the nature of regions, inclusive of their inhabitants. Its task is to investigate the distribution of phenomena in mutual dependence."

Professor Neumann of the University of Freiburg, has this year declared as follows: "General geography deals with the general laws of the distribution of every class of phenomena on the earth's surface. Special geography describes and explains the various countries in their characteristic peculiarities of land and water forms, climate, vegetation, animal life, human settlements and their conditions of organization and culture."

Mr. Mackinder, reader in geography at the University of Oxford, has, of all men in Great Britain, given most attention to the organization of geography and its relation to education. In his presidential address before the geography section of the British Association for the Advancement of Science, last year, he said: "The geographer is concerned with the *atmosphere*, the *hydrosphere*, and the surface of the *lithosphere*. His first business is to define the form or relief of the surface of the *solid* sphere and the movements or circulations within the two *fluid* spheres. The land relief conditions the circulation, and this in turn gradually changes the land relief. The circulation modifies climates, and these, together with the relief, constitute the environments of plants, animals and men. This is the main chain of the geographical argument. In the language of Richthofen, 'the earth's surface and man are the terminal links?'"

Professor de Lapparent of Paris, in his physical geography published during the present year, writes: "While ancient geography accords a preponderant place to all that concerns man, the new teaching not only discards that order of considerations, but claims to subordinate human action to the influence of nature. * * * On the one hand it embraces the precise definition, *according to form and origin*, of all the homogeneous units into which the surface of the globe can be divided. On the other hand, it inquires how these forms react upon those external physical conditions upon which all the surface activities of our planet depend, both in the mineral kingdom and in the organic world. It then completes its work by drawing a picture of the result produced by this combination of diverse elements, wherein human activity plays its large and proper part."

If it be desirable to condense all these views into a concise formula, perhaps none better can be found than this: Geography treats of the distribution of all terrestrial phenomena in mutual dependence. Or, to crowd all possible meaning into the word "science," *Geography is the science of distributions.* If this be too indefinite by reason of brevity, it may be expanded by specifying that geography is the science which deals with the mutual relations in space of relief, climate and life. This is essentially the same thought which Guyot expressed more figuratively when he said that geography views the earth as a living organism. It is only the legitimate development of an idea as old as Eratosthenes and Strabo, who aimed at the accurate location in space of every feature of the earth. Geography deals with space relations, history with time relations; therefore geography is in no sense a historical science.

The foundation of the geographic structure, or the first link in "the chain of geographic argument," is the new science of *geomorphology*, which undertakes to study the structure and origin of relief forms, much as the anatomist studies and describes the structure of the various organs of the human body. The second course in the pyramid of geography might be called *geophysiology*, because it is a study of the vital circulations which are taking place in the ocean and the air. The third course is *geobiology*, which deals with the vegetable and animal elements of the earth organism. The last and crowning block, at the apex of the pyramid, is *geoanthropology*, or the science of the relations of earth and man. Each division postulates and rests upon all the preceding divisions, and each element in turn reacts upon all the other elements. Geography lays all sciences under contribution for materials, but no other science pretends to build these materials into such a structure. Every auxiliary science contributes a certain quota of brick and stone, but geography furnishes the ground, draws the plans, and erects with enduring cement a grand and imposing temple.

The special and peculiar instrument of expression in geography is the map, because a map shows the facts of distribution better than anything else can. The series of maps prepared by Mr. Gannett to accompany the Report of the Tenth Census of the United States illustrates to perfection the special function of geography. It contains maps showing the relief of the United States; the maximum, minimum, and mean annual temperatures; the mean annual rainfall; the distribution of metals and forests; the production of grain, cotton and cattle; the density of human population, and many others. Workers in many scientific fields contributed materials, but it remained for the geographer alone to make these maps and by their means to show the remarkable relations which exist between relief, climate, products and people.

Perhaps the most striking feature of the new geography is the prominence which it gives to the study of relief. It is not content with a superficial description of plains, plateaus, mountains and valleys, but recognizes the fact that these forms of the land possess a structure and have had a history; and, above all, that they cannot be truly seen, understood or described until they are studied in the light of their origin. The scientific geographer does not admit that there are any "dead forms" in nature. The surface of the land itself is as truly undergoing a process of evolution as are the flora and fauna which inhabit it. The scientist sees as clearly as the poet that

> "The hills are shadows and they flow
> From form to form, and nothing stands;
> They melt like mist, the solid lands,
> Like clouds they shape themselves and go."

This is the new science which the geologists of the last decade have created and now hand over to the geographers; a fitting tribute from the youngest of the sciences to the oldest—the mother of all. *Physiography*, or the mutual reactions of earth and air, of relief and climate—*geomorphology*, or the science of land forms, are some of the names by which it is known; but, under whatever name, it has become the foundation of rational geography. Those who hold to the dogma that every science must be fenced off from every other science by a thought-tight barrier may protest against this invasion, as they deem it, of the field of geology. But the fact is that geomorphology is a common domain held by both sciences as tenants in entirety. The geologist can read the past history of the earth only by a study of the forms and processes now in existence. The geographer cannot understand existing forms without knowing something of their history. Both use the same material but for different purposes. As Mr. Mackinder has happily expressed it, "the geographer studies the present in the light of the past, the geologist studies the past in the light of the present." One is concerned primarily with the time relations of phenomena, the other primarily with space relations. Without a knowledge of the present the historian of the earth is helpless; without a knowledge of the past the student of the present earth is badly crippled. The twin sciences of geology and geography are indissolubly united at their common foundation, and the failure to recognize this fact is the chief cause of the deplorable state into which the prevalent geography of the schools has fallen. The influence of the new science of geomorphology upon geography is likely to prove far-reaching and favorable. Without it geography has been a pyramid resting upon its apex; a castle in the air without adequate foundation. The surface of the earth has been pictured and described superficially and without perspective. For want of depth and perspective children are still being taught that a volcano is a burning mountain, that a mountain range is a row of mountains in line, and that the Niagara gorge or Colorado Canyon was made by some great convulsion of nature. To say that all great elevations of land are mountains is as great a mistake as to say that all swimmers are fishes and all fliers birds.

The new science of geomorphology possesses the great educational merit that it can be studied in the field; and the field is everywhere, or at least wherever the natural surface of the earth can be reached. There are very few schools within easy walking distance of which cannot be found a valley and a stream—that universal concurrence of a valley and a stream which has been the despair of the geologist for a hundred years. It is easy enough to account for the stream, but the valley has been a puzzle. After trying all other explanations and finding them inadequate, the very simple conclusion has been reached that the stream has made its own valley. This idea once grasped the way is plain. A careful study of even a very small stream and its drainage basin will reveal in surprising detail the processes which have been shaping the face of the earth ever since it rose above the sea. The universal progress of weathering, transportation, corrasion, erosion and sedimentation is seen going on under the very eyes of the children. The materials of the earth-crust, its diversity of structure and the evolution in miniature of nearly every feature on the surface of the globe are displayed in endless variety. Every landscape acquires a new interest and meaning. The student obtains from his own experience a basis with which to correlate information about regions he has never seen. He has learned the alphabet in which nature has written her cuneiform inscriptions all over the face of the earth, and he can read her records. Such work as this takes geography out of the list of merely informational studies and gives it as much value for scientific training as any other science. The student who has had a taste of this will never again be content to cram facts, but will be likely to ask the sometimes awkward questions, what is the reason? how did that come to be so?

Along with this field work out of doors goes laboratory work (and every school-room can be a laboratory), consisting of actual daily observations of the sun, stars and sky, of wind, rain and snow, of temperature, humidity and pressure. The geography of the air is more difficult than that of the earth, but the teacher who knows the subject can do a great deal toward giving pupils a correct understanding of weather and climate, and can avoid the pure mythology which too many text-books contain upon this subject. The relation of plants to soil and atmosphere is within the grasp of very young pupils, and it is as easy to understand as that all animals depend upon plants for food. Having laid such a foundation, the student is prepared to see something of the crowning relation of geography—that of man to his whole physical environment; and without that foundation this relation is meaningless, because one of its elements is wanting. Herein lies the chief failure of the old geography, that it attempts the impossible. It begins at the top and builds in the air. The relation of earth to the institutional life of man is one of the most complex relations of science, and one hazards nothing in saying that not one student of geography in a thousand has had sufficient training in seeing simpler relations, or knows enough about either the earth or human institutions to see *their* relations. Geography studied in logical sequence and by scientific methods becomes in turn one of the indispensable foundations of history, sociology and political economy. It bridges the whole space between the sciences of nature and the sciences of man.

The teachers of the United States are specially fortunate in having the organization, aims, methods and spirit of the new geography clearly set before them in the report of the Conference on Geography to the Committee of Ten. If this paper accomplishes nothing more than to call renewed and serious attention to that report it will serve the writer's purpose. The whole subject is there presented with a logical power and richness of detail which are unrivaled. The key-notes of that report are *field work* and *scientific explanation*. Observation, reproduction, reasoning, are the very essence of the new geography. The report seems to have been written in the shadow of two convictions; first, that its recommendations are revolutionary; and second, that to put them into execution is, under present conditions, a matter of extreme difficulty. Of all school subjects, geography has partaken least in the recent *renaissance*. Its materials and methods are scarcely better than twenty years ago. The only natural science which forms a part of every school course is taught less scientifically than any other subject. If the writer were to speak solely from his own observation and experience, he would say that the general results of geography teaching in the grades are next to nothing except a mild dislike for the subject. Wherever the teaching is an attempted cramming of facts from the text-book, dullness and disgust are inevitable consequences, and long before the high school course is completed the facts have all evaporated. A raw country boy with only the most elementary training is a more promising student of scientific geography than the average high school graduate. The unconscious education of country life counts for much, the fact of sex counts for more. That the average boy has seen a great deal more of nature than the average girl is the natural result of their respective habits of life. The fact that nearly all the teachers of geography are women is a serious bar to the growth of better methods; because as a rule, women have had very little experience in the field. It is useless to send a class into the field with a teacher who can see nothing when she gets there.

Another serious difficulty in the way of better methods of teaching is the want of opportunity for the teacher to obtain special training in geography. Very few colleges and universities in this country recognize the existence of geography as a distinct science, but there are some notable exceptions to the rule. Harvard heads the list, where the department of physical geography under the direction of Professor Davis has become a veritable fountain head from which good influence has penetrated in every direction. Cornell, Chicago, Princeton, Yale, Rochester, Leland Stanford, Oberlin and Colgate are good secondary centers. As for the Normal schools, they are, with few exceptions, still in the dark ages upon this subject, and apparently likely to remain strongholds of conservatism.

The outlook for progress in this direction is not wholly without encouragement. Harvard, Chicago and other institutions offer summer courses in geography which are fairly well attended. A course of five lectures on physical geography by Professor Brigham of Colgate at the Cook County Institute last summer was attended and appreciated by four hundred teachers. That the new geography is making some progress may be judged from the text-books. Guyot's and Houston's contain about one page each upon the subject of geomorphology, the Eclectic one hundred pages, Tarr's two hundred pages, while the French geography of de Lapparent, the most advanced along the new lines, devotes three hundred pages to this subject. Good literature and personal training are becoming more available every year, and no teacher in search of them need fail to find both. It is here that school officers and superintendents can do more than any other influence. Professor Davis of Harvard, in reply to a request for a teacher writes: "It is only occasionally that a student here takes enough geography to gain strong recommendation for teaching, because employers, thus far, have given no weight to geographical preparation for a geographical teacher; anyone might do that sort of work." If employers of teachers would require a preparation in geography equivalent to that required in literature or history, a demand for such preparation would be created, and higher schools would soon furnish a supply equal to the demand.

This paper has aimed to set forth the following thoughts concerning the new geography:

1. Its philosophy is not teleological, but evolutionary. It is no longer anthropocentric, but geocentric.

2. The new geography is scientific and rational. It studies not only facts (which are stupid things), but the relations between facts.

3. The new geography has been enriched by the addition at the bottom

of the new science of geomorphology, and is thus brought into close alliance with geology.

4. The new geography forms a connected chain between the purely natural sciences and the humanities; but being preponderatingly a natural science it must adopt the scientific or laboratory methods of study and teaching.

5. Thus the new geography becomes able to give, not only information, but scientific training; the ability to discover facts and to see their relations. It converts geography from a lifeless bore to a living interest, from a dead grind to a delightful activity. It takes it out of the list of memory or "useful knowledge" studies, and plants it in the quickening current of modern scientific thought.

6. It is only when built upon "the solid ground of nature" and inspired by the scientific spirit that geography can hope to solve the problem of Ritter and Buckle: the problem of the relation of man to his physical environment, and thus, become in fact, the physical basis of history and sociology.

7. Special means must be adopted to prepare teachers for this kind of work. On account of lack of special training and lack of facilities for obtaining it, educational progress in this direction will be slow; but the new geography has come to stay, and teachers and school officers will do well to recognize and welcome it.

III.—THE GENERAL GEOGRAPHY OF INDIANA

CHARLES R. DRYER

POSITION AND BOUNDARY

Indiana is one of the North Central states, situated in what is sometimes called the Middle West, between the upper Great Lakes and the Ohio, and mostly in the Mississippi basin. The central parallel of the United States, the 39th, crosses its southern third and it is included between 37° 41′ and 41° 46′ north latitude, and between 84° 44′ and 88° 6′ west longitude. It is bounded on the north by the parallel which is ten miles north of the southern extremity of Lake Michigan; on the east by the meridian of the mouth of the Great Miami river; on the south by the Ohio; and on the west by the Wabash river and the meridian of Vincennes. Its extreme length is 250 miles, its average width 145 miles, its area, 36,350 square miles.

ELEVATION

According to Powell's division of the United States into physiographic regions,* Indiana lies mostly on the Ice Plains, but includes a small portion of the Lake Plains on the north, and of the Alleghany Plateau on the southeast. The highest land in the state, in Southern Randolph county, is 1,285 feet above tide; the lowest, at the southwest corner, is 313 feet. The area above 1,000 feet comprises 2,850 square miles in three tracts: (1) an irregular area around the headwaters of the Whitewater river in Union, Wayne, Randolph, Delaware, Henry, Rush, Decatur, Franklin and Ripley counties; (2) a narrow crescentric ridge in Brown county; (3) a considerable area in Steuben, DeKalb, Noble, and Lagrange counties. Isolated peaks rise in Brown county to 1,172 feet, and in Steuben to 1,200 feet. The area between 500 and 1,000 feet in elevation is 28,800 square miles, and that below 500 feet is 4,700 square miles. The average elevation of the state is 700 feet.

---
*National Geographic Monographs, No. 3.

Geological Structure

The rocks of Indiana are all sedimentary, and consist of a series of shales, sandstones and limestones laid down upon the bed of a shallow ocean off the shore of a land area which lay to the eastward. These strata are shown by borings to be more than 3,000 feet thick. They have never been compressed, folded or violently disturbed; but have been gently lifted into a very flat arch, the crest of which extends from Union county to Lake county. From the crest of the arch the strata dip gently to the northeast and southwest, the slope in the latter direction being about twenty feet to the mile.* The oldest rocks exposed are the Hudson river shales, in the southeast; the youngest are the Carboniferous, along the west side.

Physical History

Indiana has been a land surface since the close of the coal period. Subjected during those millions of years to weather and stream erosion, it was maturely dissected into a complex network of valleys, inter-stream ridges, and isolated buttes. Over this surface the great Laurentide glacier repeatedly passed, extending once as far as the glacial boundary shown on the map, and again to the "Wisconsin" boundary.† Its effect was to grind down and smooth off the hills, to fill up the valleys, and to leave the surface plastered over with a great mass of loose material, much of which was brought from northern regions. Since the final disappearance of the ice the streams have partially cleared out a few of the old valleys and have begun to cut a system of new ones in the surface of the drift, but this cycle of erosion is still in its infancy. Thus, the greater part of Indiana is a plain of glacial accumulation, of very recent origin, and too young to have developed more than rudimentary features.

Physiographic Regions

The most striking physical contrast in Indiana is that between the glaciated and unglaciated areas. A comparison of the topographic map with that showing the revised glacial boundary brings out this contrast sharply. North of the limit of drift the contour lines‡ run in large

---

* See the excellent sections of Professor Cubberly, showing the structural features of Indiana, in 18th Report of State Geologist, p. 219.

† See Maps, pp. 26, 28.

‡ Contour lines are lines of equal elevation which run across the country, each everywhere at the same height above the sea. The shore of the ocean is the basal contour line, and if the sea level should rise a hundred feet it would mark a new contour line at that level. Where contour lines are far apart the slopes are gentle and the surface comparatively smooth; where they are close together the slopes are steep and the surface rough and broken. The contour lines on the topographic map of Indiana are general and approximate only. Fuller and more accurate surveys are necessary before they can be drawn with exactness and detail.

curves and are far apart, showing the general smoothness and monotony of the surface. South of the glacial boundary the lines are crowded and extremely tortuous, showing a surface much cut up. The limit of drift encloses and fits this area of broken surface as a man's coat fits his shoulders.

The Ohio Slope.—That portion of the state which slopes directly to the Ohio, including the driftless area and the southeastern part of the drift plain, is a region of deep, narrow valleys, bounded by precipitous bluffs, and separated by sharp, irregular divides. Isolated knobs and buttes are numerous; the crests and summits are from 300 to 500 feet above the valley bottoms. The streams are rapid and broken by frequent cataracts. All open out into the Ohio Valley, a trench from one to six miles wide, 400 feet deep and bounded by steep bluffs.

The Central Plain.—North of an irregular line extending in a general direction from Richmond to Terre Haute, and south of the westward flowing portion of the Wabash from Fort Wayne to Attica, the topography is that of an almost featureless drift plain. It is traversed by numerous morainic ridges, but they are low and inconspicuous. The traveler may ride upon the railway train for hours without seeing a greater elevation than a hay stack or a pile of saw dust. The divides are flat and sometimes swampy, the streams muddy and sluggish. The valleys begin on the uplands as scarcely perceptible grooves in the compact boulder clay, widen much more rapidly than they deepen, and seldom reach down to the rock floor.

The Northern Plain.—The portion of the drift plain north of the Wabash river is more varied than the central plain, and comprises several regions which differ materially in character. A small area around the head of Lake Michigan is occupied by sand ridges and dunes, partly due to a former extension of the lake and partly to present wind action. Some of the drifting dunes are more than 100 feet high. This region is separated by a belt of morainic hills from the basin of the Kankakee, which contains the most extensive marshes and prairies in the state. This region also is traversed by numerous low ridges of sand, the origin and character of which are not yet well understood. Many of its features are probably due to the fact that during the retreat of the ice-sheet it was temporarily occupied by a glacial lake, which received the wash from the high moraines to the eastward. Northeastern Indiana is *the region of high moraines*, and has a strongly marked character of its own. A massive ridge of drift, twenty-five miles wide, 100 miles long, and from 200–500 feet thick, extends from Steuben county to Cass county, and is joined by several smaller branches from the northwest. This is the joint moraine of the Erie and Saginaw lobes of the Laurentide glacier. Much of its surface is extremely irregular, presenting a succession of rounded domes, conical peaks, and winding ridges, with hollows of corresponding shape between, which are occupied by innumerable lakes and marshes; the highest points are 100-300 feet above the level intermorainic intervals. A large proportion of the material is sand and gravel. A small area in eastern Allen county is a part of the bed of the glacial Lake Maumee.

Drainage

The general slope of Indiana is to the southwest, as indicated by the course of the Wabash river and its tributaries, which drain two-thirds of the state. Of the remaining third, one-half is drained directly to the Ohio and one-half to Lakes Erie and Michigan, and to the Mississippi through the Illinois.

The Wabash River is the great artery of Indiana, which it traverses for more than 400 miles. The fall is quite uniformly about eighteen inches per mile. Its current is gentle and unbroken by notable rapids or falls. Its valley is quite varied in character. Above Huntington it is a young valley, without bluffs, terraces or flood plain. Below Huntington, it once carried the drainage of the upper Maumee basin, and is nowhere less than a mile wide as far down as Attica. Below that point its width varies from two to six miles. The original valley has been largely filled with drift, which the present river has been unable to clear out. It winds between extensive terraces of gravel, which border it at various elevations, and flows at a level from fifty to one hundred feet above the original rock floor. Below Terre Haute, the wide flood plain, ox-bow bends and bayous give it a character similar to that of the lower Mississippi. The upper tributaries as far down as Lafayette are post-glacial streams in drift valleys, whose courses are largely determined by the trend of the moraines. Below that point the smaller tributaries enter the river through picturesque sandstone gorges.

White River, the largest tributary of the Wabash and rivaling it in volume of discharge, is a much more varied and complex stream. The larger West Fork rises at the summit level of the state in Randolph county. In its upper course it is moraine-guided, like the upper tributaries of the Wabash, and presents the same characters as the other streams of the central plain. In Morgan county it assumes a different aspect, and thence to its mouth flows through a valley from one to three miles wide, 100 to 300 feet deep, bordered by wide bottoms. The East Fork rises on the same elevation as the West, but reaches its destination by a more tortuous course. Although its length is increased and its slope decreased by its numerous meanders, it is still a swift stream. Both forks of White river suffered many disturbances during the glacial period, which have not yet been studied in detail, but are obvious from the varying character of their valleys and from the terraces which border them at all heights up to 300 feet.

The Whitewater River, takes the shortest course of all from the summit level to the Ohio, and its average fall is about seven feet to the mile. At Richmond it has cut a narrow gorge into the soft shales 100 feet deep. In strongest contrast with this and the other rivers of the Ohio Slope is the *Kankakee*, which winds through wide marshes with a scarcely perceptible current, and without definite banks. Its basin, however, is sufficiently elevated to render good drainage possible by the construction of the requisite ditches, and much has already been done to that end.

Physiographic Features

Many important land forms are wanting in Indiana. There are no mountains, no valleys formed by upheaval or subsidence, no volcanoes or volcanic rocks except foreign fragments brought by the ice-sheet, no features due to disturbance of the earth crust except the rock foundations of the state itself.

Plains.—As already indicated, the greater part of Indiana is a *plain of accumulation*; the surface of a sheet of glacial drift which varies in thickness from a few feet to 500 or more. The average thickness is more than 100 feet. It consists chiefly of a mass of clay containing more or less gravel and boulders—the *till* or boulder clay of the geologists. This is locally varied by heaps, ridges, sheets and pockets of sand and gravel, and in the southern part of the state is overlain by a peculiar fine silt called *loess*. The boulder clay is the grist of the glacial mill, and is composed of a very intimate and heterogeneous mixture of native and foreign materials, containing fragments of almost every known mineral and rock. The large fragments, or boulders, are widely distributed, and of every size up to thirty feet in diameter. They are nearly all igneous or metamorphic in character and can be traced back to their origin in the Canadian highlands north of the Great Lakes.

The driftless area is a *plain of degradation*, formed by the removal of the original rock surface to an unknown depth, and now represented by the summits of the flat and even-topped divides, ridges and hills.

Hills.—On the northern plain occur numerous *hills of accumulation* forming the great morainic belts, the result of excessive dumping and heaping up of drift along the margins and between the lobes of the melting ice-sheet. The most impressive examples a refound in Steuben, Lagrange, Noble and Kosciusko counties, where they attain a height of 200 feet or more, and are as steep and sharp as the materials will lie. Their peculiar forms and tumultuous arrangement give a striking and picturesque character to the landscape.

The Ohio Slope is studded all over with *hills of degradation*—blocks and fragments of the original plain left by the cutting out of the valleys between them. Some are broad and flat-topped, some narrow, crooked and level-crested, some sharp or rounded, isolated knobs or buttes. These are very conspicuous in the counties of Greene, Daviess, Martin, Crawford, Orange, Washington and Jackson, but attain their greatest development in Floyd, Clarke and Scott, where the Silver Hills and Guinea Hills rise to 400 and 500 feet above the valley bottoms. In Brown county the knob topography attains the highest absolute elevation in Weed Patch Hill, and the surrounding region is so rugged as to have gained the title of the "Switzerland of Indiana."

In Benton county, Mounts Nebo and Gilboa are isolated masses of rock projecting above the general level of the plain, and are probably entitled to the name of *monadnocks*.

Moraines.—In addition to the massive and rugged moraine belts already described, there are many morainic ridges of gentle slope and smooth profile, "like dead waves upon the surface of the ocean," conspicuous only upon the map by their influence upon streams. Those which extend along the right bank of the St. Mary's, upper Wabash, Salamonie, Mississinewa, and upper White rivers are typical examples. The southernmost moraine in the state, which enters Vigo, Vermillion and Parke counties from Illinois, is composed largely of a series of broad, low mounds, irregularly disposed upon the plain. In this connection should be mentioned the form of moraine known as *boulder belts*—long, narrow, curving strips of country thickly covered with large boulders. These occur in the counties of Jasper, Newton, Benton, Warren, Tippecanoe, Boone, Clinton, Hendricks, Johnson, Shelby, Rush, Henry, Randolph, Wayne, Whitley and Huntington.

Kames and Eskers.—These are deposits of sand and gravel laid down by strong streams of water which flowed from the edge of the melting ice-sheet. Kames are irregular ridges and mounds, having a general direction at right angles to the direction of ice movement, and are found in connection with the massive moraines. Eskers, or "serpent kames" are long, winding ridges of sand and gravel, parallel to the direction of ice movement, and generally extending down a valley of glacial drainage. They mark the course of streams which flowed in sub-glacial tunnels. The valley of Turkey creek, in southwestern Noble county, the Erie-Wabash channel southwest from Fort Wayne, and the whole course of the "Collett Glacial River," from Delaware and Madison to Decatur and Bartholomew counties present numerous examples. There are probably many more in the state still unreported.

Closely related to these are *sand and gravel streams, plains and overwash aprons*, in which the material is spread out over broader areas. Northern

Steuben, Northwestern Whitley and Central Bartholomew counties contain good examples, a few out of a probably large number in the state.

Dunes and Beach Ridges.—These are hills and ridges of sand or gravel, either blown up by the wind, or built up by the waves along the shores of lakes now withdrawn. The region around the head of Lake Michigan, the Kankakee basin, and the Maumee Lake basin east of Fort Wayne afford fields for more extensive study of these forms.

Valleys.—As before stated, all the valleys of Indiana are the result of stream erosion; most of them by the streams which now occupy them. During the glacial period, however, the streams generally carried much more water than at present.

Gorges, Ravines and Canyons are deep, narrow valleys with precipitous walls. They exist in great number and variety throughout the Ohio Slope, occurring along the Whitewater, White and Ohio rivers, and all their tributaries. The eastern tributaries of the Wabash in Fountain and Parke counties flow through very beautiful canyons cut in massive sandstone, often with overhanging walls which, at "The Shades of Death," reach a height of 250 feet.

In valleys of this character *rapids and falls* are very numerous. They occur upon nearly every stream emptying into the Ohio, and vary in height from a few feet to sixty or eighty. Clifty Falls in Jefferson county and Cataract in Owen county are among the most famous.

All the streams flowing from the glacial area, had their valleys flooded with glacial waters, and choked with glacial debris. The effects of this are shown by the extensive *terraces* of sand and gravel which border their present channels, and mark the heights at which they were once able to deposit sediment. Between these terraces there are often broad "bottoms" or *flood plains* which furnish the best corn lands in the world.

Glacial Drainage Channels.—During the melting of the ice-sheet the waters found escape by numerous channels which are not now occupied by any large or continuous stream. A very notable one is the Erie-Wabash channel, which carried the waters of the glacial Lake Maumee from Fort Wayne into the Wabash at Huntington. The largest in the state gathered the water from numerous channels in Jay, Grant, Blackford, Randolph, Delaware, Madison and Henry counties into one great stream, which flowed southward through Hancock, Shelby, Bartholomew, Jennings, Jackson, Scott and Clark to the Ohio at Jeffersonville. In its middle course its valley is forty miles wide and 400 feet deep, narrowing to five miles near its mouth. It has been named the "Collett Glacial River."

Lakes.—The surface of the northern plain is peppered with small lakes which occupy irregular depressions in the surface of the drift, and are especially characteristic of the massive moraines. The whole number cannot be less than 1,000. The largest, Turkey Lake in Kosciusko county, has an area of five and a half square miles.

Marshes and Swamps.—These exceed the lakes in number and extent. The smaller ones are the basins of former lakes which have been filled up with sediment and vegetation. The largest are in the Kankakee basin, and are the remaining vestiges of a glacial lake. Everywhere over the central plain the divides are too flat and the slopes too gentle for good drainage, and marshes abound. These, however, have been largely drained by ditches.

Sinkholes and Caves.—Extending from Harrison, Crawford and Clarke counties to Putnam is a belt of limestone which is honey-combed by underground streams producing a great variety of sinkholes, caves and "lost rivers." The sinkholes are basin-like depressions ten to fifty feet deep, and thirty to three hundred feet in diameter, with an opening at the center which leads to some underground passage. In some cases a stream drops into this hole out of sight and emerges again upon the surface many miles away. If the opening has become clogged the basin holds a pool of clear water. Many of the underground passages have been wholly or partially abandoned by the streams which made them, and can be followed great distances. Wyandotte cave in Crawford county has been explored a distance of twenty-three miles, and rivals in extent and beauty the Mammoth Cave of Kentucky. In Harrison county the rocks are so fissured as to render wells uncertain. This county contains Ripperden, Harrison and Grassy valleys, which are closed amphitheaters of three to ten square miles in area and 200 to 400 feet deep, and formed by the falling-in of the roof of subteranean caverns.

CLIMATE AND VEGETATION

The map shows the rainfall and mean temperature for 1896, which was very nearly a normal year. The mean temperature for January varies from 25° in the north to 33° in the south, for July from 72° to 77°. The absolute extremes of temperature for the state and year are 103° and —22°. The number of days in the year with average temperature below freezing is ninety in the north and twenty in the south. The changes of temperature are frequent except in summer, when a period of two or three months of uniformly warm, clear weather often occurs. The mean rainfall is quite variable from year to year, ranges from thirty-five inches in the north to forty-five in the south, and is well distributed throughout the year with a slight excess in spring. The average annual snowfall in the north is forty inches, in the south fifteen. The prevailing winds are from the southwest, and the average wind velocity seven to nine miles per hour. Thunderstorms and tornadoes are frequent.

THE GENERAL GEOGRAPHY OF INDIANA

Contrary to the statements made in many books, Indiana is not a prairie state. An area estimated to comprise one-eighth of the whole, situated, except a few isolated patches in the northwestern part, is marsh and upland prairie. The remainder of the state was originally covered by a heavy growth of oak, walnut, beech, maple and other hardwood timber, with sycamore and poplar near the streams, and a little pine along the Ohio slope. No region in the United States could show finer specimens, or a greater number of individuals and species of forest trees than the lower Wabash Valley. The same region is said to be the original habitat of the bluegrass which has made Indiana and Kentucky pastures so famous.

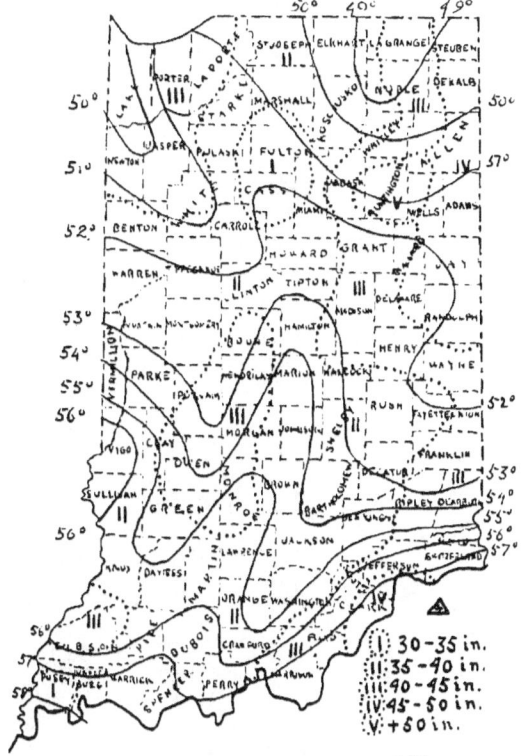

RAINFALL AND MEAN TEMPERATURE FOR 1896.

Resources

Mineral.—As shown upon the map, an area in the southwestern part of the state, comprising 7,000 square miles, is underlain by numerous seams of bituminous and block coal, which is mined to the extent of four million tons yearly. The natural gas field in the East Central part, an area of 2,500 square miles, furnishes gas to the value of $1,500,000 annually, which is used by numerous manufacturers in the field, and is piped to all the neighboring cities and to Chicago. Indiana is second only to Pennsylvania as a gas producing state. On the northern border of the gas field is a small but rich and growing petroleum field. A nar-

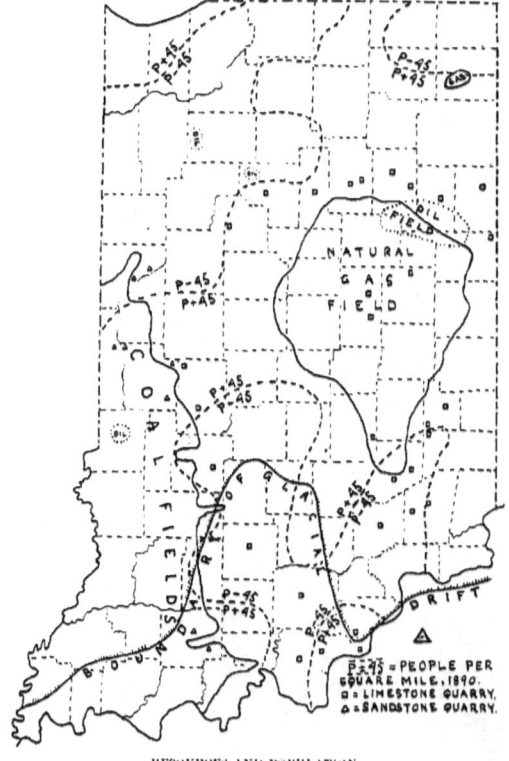

RESOURCES AND POPULATION.

row belt extending from Washington to Putnam county furnishes the best limestone in the world for building purposes. Clays and shales, suitable for brick and tile making, occur in nearly every county. Lawrence, Martin and Owen counties contain deposits of kaolin of sufficiently high grade for the manufacture of fine pottery. Floyd and Parke counties furnish a good quality of glass-sand.

Agricultural.—Agriculture has been, and probably will always remain, the foundation of Indiana's prosperity. The glacial drift is a very productive and permanent soil, especially for the cereal grains. The level tracts of boulder clay require under-draining to obtain the best results. The alluvial soils of the bottoms and terraces along the larger streams cannot be surpassed for corn production. The bluffs, knobs and hills of the driftless area along the Ohio have been found favorable for the growing of apples, peaches, grapes and other fruits. The marshes of the Northern Plain, when properly drained, yield large crops of hay, corn and celery. Although among the states Indiana ranks thirty-fourth in area, she was in 1889 seventh in the production of cereals and of corn, and fourth in the production of wheat.

Settlement and Population

Settlement from the Middle and Southern states began along the Ohio early in the present century and extended northward. Forty years later a stream of New England and New York people came into the northern part. The total population in 1890 was 2,192,404, of which only 18 per cent. live in cities, and less than seven per cent. are foreign-born—chiefly German. The map shows how closely the distribution of population corresponds to physical conditions, the areas of relatively sparse population including (1) most of the driftless area and the rugged and broken region of the Ohio slope, except the coal fields and best fruit growing region; (2) the prairies and marshes of the Kankakee basin, and (3) the roughest portion of the high moraine in the northeast. The influence of Chicago is shown in the northwest corner by the presence of a denser population in a region physically unfavorable. A closer analysis would probably show an area of excess in the manufacturing districts of the gas field.

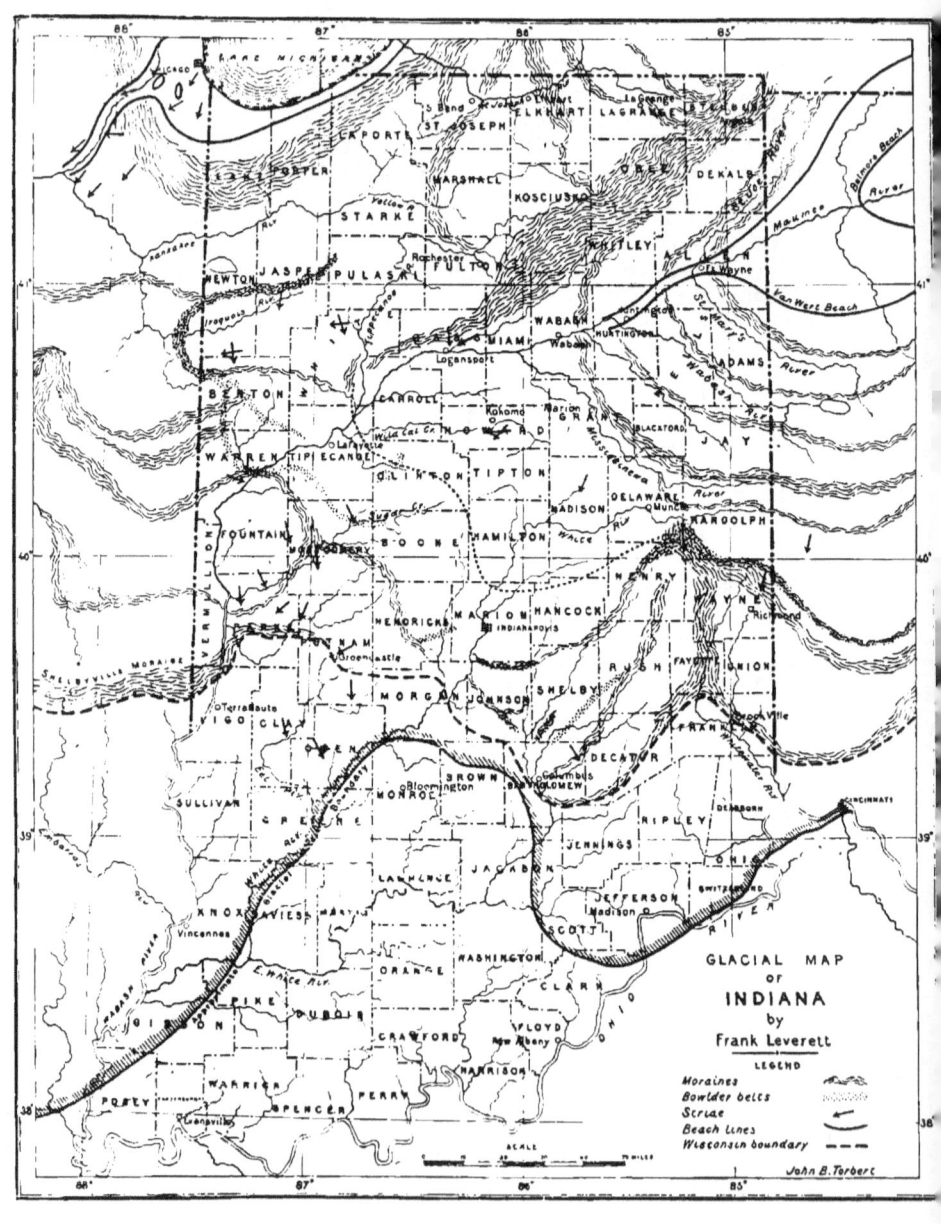

IV.—THE GLACIAL DEPOSITS OF INDIANA *

FRANK LEVERETT, F. G. S. A. UNITED STATES GEOLOGICAL SURVEY

INTRODUCTORY

In Indiana, the glacial deposits and scorings have been recognized from the earliest days of settlement; indeed, it is in this state that we find about the first recognition in America of the boulders as erratics and of striæ as products of ice action. So long ago as 1828, granite and other rocks of distant derivation were observed by geologists near New Harmony, in the southwestern part of the state.† At nearly as early a date (1842), striæ were noted near Richmond, in the eastern part of the state.‡

Notwithstanding the early date at which observations of glacial action began, very little attention was given to the drift, here or elsewhere, until within the past twenty years. It was commonly passed over in geological reports much as the soil is even to-day, with some casual remark concerning its presence in great or small amount. Within the past twenty years interest in these deposits, because of the varied history which they reveal, has been so aroused, that many geologists, both in America and Europe are making a systematic study of them.

In Indiana these deposits are engaging the attention of both the State and the United States Survey. The study of general features and a comparative study of the drift of Indiana and neighboring states has been undertaken by the United States Survey, while the detailed examination of deposits has been entered upon by the State Survey. Professor T. C. Chamberlin has superintended the United States Survey work and has himself spent considerable time in Northern and Western Indiana. Under his direction Professor G. F. Wright and Professor J. C. Branner, have investigated the glacial boundary; Professor L. C. Wooster has studied the district north of the Kankakee, and the writer has made a

---

Note Concerning the Glacial Boundary.—Further study during the season of 1896 by Mr. Leverett and Messrs. Ashley and Siebenthal of the Indiana Geological Survey has determined the occurrence of glacial drift as far south as the revised line shown on map on page 26.

†See Geology of Indiana, 1878, pp. 105-106.

‡See Amer. Jour. Sci., Vol. XLIV, 1842-3, pp. 281-313.

reconnoissance of nearly all the drift-covered part of the state. Professor Wright's results have already been published in the United States Geological Survey bulletin, No. 58, issued in 1890. Professor Chamberlin's earlier results are set forth in his paper on the "Terminal Moraine," in the Third Annual Report of the United States Geological Survey for 1881–82. The later results of his studies and those of Professors Branner, Wooster and the writer, are largely unpublished. Through the courtesy of Professor Chamberlin the writer is permitted to set forth some of the leading results in this paper.

The work of the State Survey has not been uniform. Portions of it have been less detailed than that of the United States Survey, while other portions have been carried into greater detail. Probably the most detailed and careful study of any considerable area is that made by Dr. C. R. Dryer in the northeastern part of the state.* An examination of the reports of the Indiana Geological Survey will serve to set forth these differences and to show the importance of extending the detailed study of glacial deposits over all the glaciated portions of the state. Such a study probably can be carried on to the best advantage under the organization of a State Survey. But independent workers can do much to throw light on these deposits by collecting the records of well-borings and by careful notes taken at natural or artificial exposures.

Before entering upon the discussion of the Indiana drift a few words of explanation seem necessary concerning the material of the drift, and concerning the gathering grounds of the ice which overspread this region.

MATERIALS OF THE DRIFT

It is quite a prevalent idea that the boulders which strew the surface of the glaciated districts and which have suffered transportation from distant regions, constitute the most impressive evidence of ice action. It seems by many not to be understood that the thick deposits of stony clay with associated beds of sand and gravel which blanket the North Central States to a depth of 100, 200, and occasionally 500 feet, are also due to ice transportation. Over a large part of the country from the Dakotas eastward to the Appalachian ranges, these deposits are so thick that ordinary wells fail to reach their bottom, and many of the valleys of the large streams are formed entirely in them. The boulders in reality constitute but an insignificant portion, for probably ninety-five per cent. of the drift of these states consists of minute rock fragments and sand and clay, and of the remaining five per cent. only a small part is made up of large blocks of distant derivation; i. e., of boulders proper.

An examination of rocks in the drift mass will usually disclose a large percentage of material which has not been transported far, but there is

---
* See sixteenth, seventeenth and eighteenth reports of State Geologist.

usually to be found a sprinkling of rocks from distant localities. Let the reader select some space, say a square yard, in a gravel pit or other exposure and set about classifying the several kinds of rocks represented, and he will ascertain the relative amount of local and distant material.

In its bedding the drift displays great irregularity. In general, it consists of a confused mass of angular, semi-angular, and well-rounded stones imbedded in a matrix of sandy clay. This confused mass was named *till* by Scottish geologists, and this term has been adopted by American geologists. By some it is called boulder clay, because of its containing boulders. With the till one can find, in many exposures, beds or pockets of sand and gravel. These beds in some cases comprise the entire section, but they are usually subordinate to the till.

In some parts of the glaciated districts the till constitutes the lower part of the drift, while the sand and gravel lie mainly near the surface. In Indiana such a relationship does not prevail over wide areas. The drift deposits of this state are unusually varied in the arrangements of till sheets, gravel beds and sand beds; what is true of one township may find no application in a neighboring one.

Farther on we shall discuss the evidence upon which is based the conclusion that there are in Indiana drift-sheets differing widely in age.

GLACIATED ROCK SURFACES

The peculiar appearances presented by rock surfaces which have been abraded by the ice-sheet are usually of such a striking nature as to arrest the attention of untrained as well as of trained observers. These surfaces differ somewhat from place to place but still have a characteristic appearance. They present, usually, a series of parallel, or but slightly divergent lines or grooves, varying in size from faint scratches as fine as a hair, to broad, shallow grooves an inch or two, and occasionally several inches in width. Between the grooves the rock has usually been scoured down to a plane surface. The striæ indicate, as a rule, the general course of ice-movement and with few exceptions point toward the margin presented by the ice-sheet at the time they were formed.

As the ice-sheet was subject at times to excessive wastage, if not to complete destruction, followed by readvance in which some shifting of movement occurred, we find the striæ showing some interesting variations in neighboring localities. Some of the best illustrations in America are to be found in Western Indiana and these are discussed farther on.

THE GLACIAL GATHERING GROUNDS

On the glacial map of North America are shown the extent of glaciation, and the several main centers of dispersion; viz., the Cordilleran, Keewatin, Labrador and Greenland. The glaciated districts in North

America are estimated to cover 4,000,000 square miles. It is doubtful, however, if this entire area was covered by the ice-sheet at any one time. Dr. G. M. Dawson, director of the Canadian Geological Survey, has found evidence that the Cordilleran ice-field overspread the Rocky Mountains and extended some distance to the east and then withdrew before the Keewatin ice-sheet had reached that region.*

Similarly the Keewatin ice-sheet culminated and withdrew from its southern limits (in Missouri and Iowa) before the Labrador ice-field had reached its extreme western limits. The writer has found that the Labrador movement extended into Southeastern Iowa at a date considerably later than the time when the Keewatin ice-sheet withdrew; there being a soil and other evidences of an interval found on the surface of the Keewatin drift and under the drift of the Labrador sheet. It should be understood, however, that the reduction in size of the Cordilleran and Keewatin sheets at the time of the culmination of the Labrador sheet, may have amounted to but a small percentage of the area which they had covered.

Greenland is now ice-covered while districts to the west which have been ice-covered are nearly free from glaciers. The continuation of glaciation there parallels the observations in the fields to the west and adds to the weight of these observations in indicating a progressive culmination of the ice-sheet from west to east.

Aside from the four main gathering grounds there appear to have been minor gathering grounds in the extreme east on New Brunswick and on Nova Scotia as indicated by Mr. Robert Chalmers in his paper in the Annual Report of the Canadian Survey for 1894. There were also small ice-fields on the Rocky and Sierra Nevada Mountains in the Western United States, as described many years ago by King, Whitney and others.

The Glacial Succession in Indiana

First Ice Invasion.—This state was invaded by ice which had as its center of dispersion the elevated districts to the east and south of Hudson Bay. There was a movement from the region north of Lake Huron in a course west of south over the Lake Michigan basin, Illinois and Western Indiana. There was also a southward movement from the same region across Lakes Huron and Erie, Western Ohio and Eastern Indiana. It is not known whether these movements were independent and of different dates or whether there was simply a radiation in movement of a single ice accumulation. It should not be taken for granted that even within the state of Indiana the ice-sheet was occupying the glacial boundary completely at any one time.

Bulletin of the Geol. Soc'y of America, Vol. VII, pp. 31-66, November, 1895.

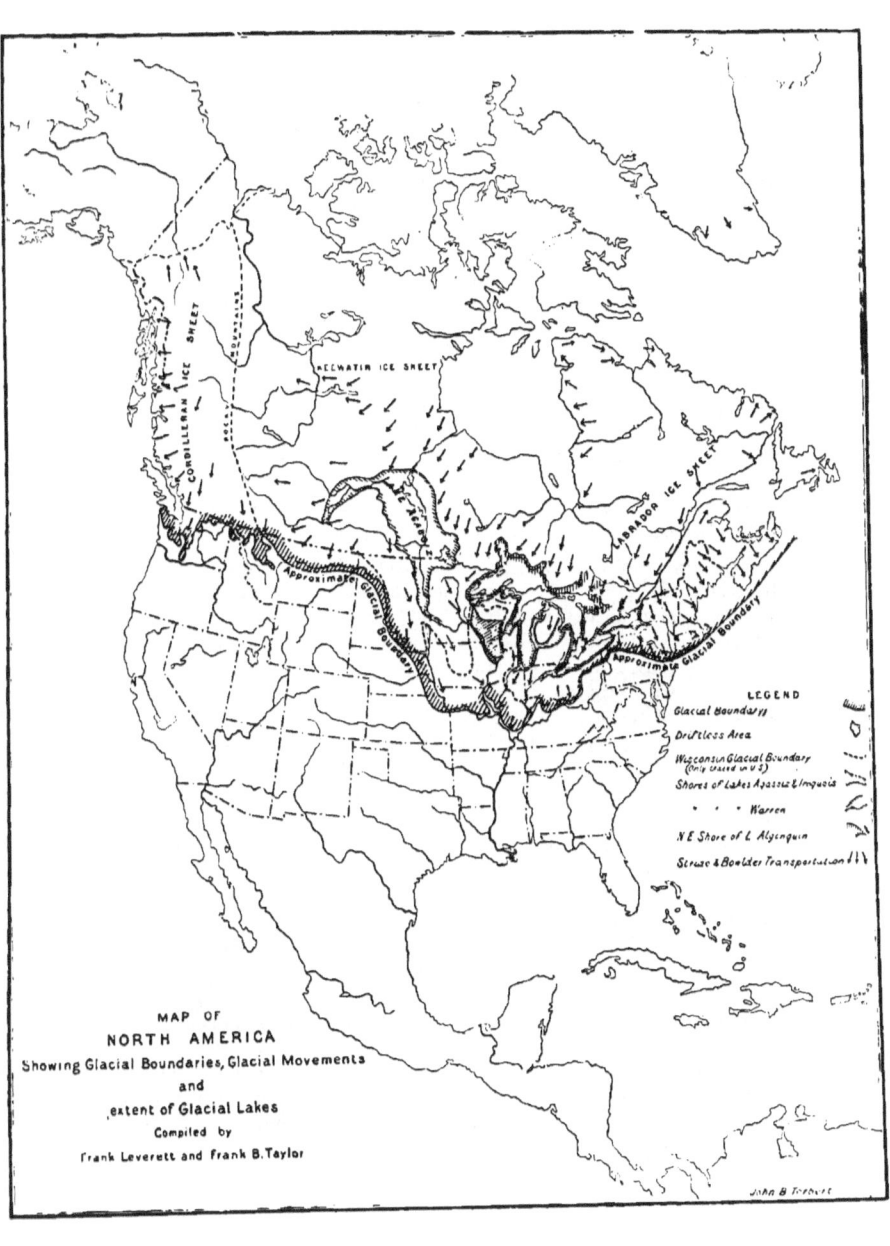

The ice deposited but little drift near its extreme limits, either in Indiana or the states to the west. There is not, as a rule, a well defined ridge or thick belt of drift along the glacial boundary, such as characterizes the southern limit of some of the later drift-sheets, though occasional ridging of drift is to be seen, as in Chestnut Ridge in Jackson county* and a similar ridge in Southern Morgan county. The boundary of the drift in Indiana is usually so vague and ill defined that it is only approximately known.

If we may judge of the deposit over the state from the outlying portions, south of deposits made by later invasions, the deposits of the first invasion are of much less volume than those of later invasions. They appear to include not more than 30 of the 130 feet which the writer estimates the state to carry. In the portion of the state which was glaciated but once the thickness is usually less than 25 feet, but filled valleys will probably give it an average somewhat above that amount. What is true of the drift of the earliest invasion in Southern Indiana is true also of the same drift of Southern Illinois and Southwestern Ohio. This invasion seems, therefore, to be quite widely characterized by a lighter deposition than that of the later invasions.

First Interglacial Interval.—After reaching the line marked by the glacial boundary, the ice melted away and left the drift exposed to atmospheric agencies. How far to the north the land became uncovered is not known. At this time a black soil was formed, which is now concealed beneath deposits of silt, termed loess, in Southern Indiana, and beneath later deposits of till in the northern portion of the state. This soil is found at the base of the loess at various points over the southern portions of the state, but is best developed on flat tracts. It may be seen beneath the loess in the flat districts east and south of Terre Haute at a depth of from six to eight feet. The vegetable matter appears to have accumulated there just as it does on the present surface of poorly drained tracts in northern latitudes, where decay is slower than accumulation. In Western Indiana, from Parke and Vermillion counties northward, the soil is found below a later sheet of till at depths varying from twenty feet up to one hundred feet or more. Numerous references to the soil below till in this portion of the state are to be found in the "Indiana Geological Reports," it has not been observed in Eastern Indiana, so far as the writer is aware, but may be present, for few valleys there reach low enough to expose it. It seems not to be so conspicuous, however, as in Western Indiana, otherwise it would have been brought to notice in well-borings.

No conclusions have been reached concerning the length of time involved in the formation of this soil. The land at that time seems to have been so low or so flat in Indiana, that drainage lines were not well

---

*Geology of Indiana, 1874, pp. 56-57.

developed in the drift surface, and we are thus deprived of one important means of estimating the work accomplished.

Main Loess Depositing Stage.—Loess is a term applied to a fine-grained yellowish silt or loam, which overspreads the southern portion of the glacial drift in North America, and extends thence southward on the borders of the Mississippi Valley to the shores of the Gulf of Mexico. The term was originally applied to deposits of this character on the Rhine, which have very extensive development in the German lowlands and bordering districts in Northern Europe. Microscopical analysis shows it to consist principally of quartz grains, but it usually has a variety of other minerals such as occur in the glacial drift. It is apparently derived from the drift, either by the action of water or wind. In many places, especially on the borders of the large valleys, the loess is charged with calcareous matter which partially cements it. When excavations are made in it the banks will stand for years, and will retain inscriptions nearly as well as the more consolidated rock formations. It has a strong tendency to vertical cleavage, and usually presents nearly perpendicular banks on the borders of streams which erode it. It often contains concretions or irregular nodules of lime and of iron and manganese oxides. It is also often highly fossiliferous. The fossils are usually land and fresh-water mollusks, but occasionally insects and bones of mammals are found.

The deposit appears to be mainly of one stage in the glacial period, and has been definitely correlated by Mr. W. J. McGee with an ice invasion which followed the interglacial stage just discussed.* In the region which Mr. McGee studied, in Northeastern Iowa, it connects on the north with a sheet of till called by him the upper till, and afterwards named by Professor Chamberlin, the Iowan Drift-Sheet. The writer has visited that region and fully concurs with Mr. McGee's opinion. This drift-sheet has not been recognized in Indiana, for if present it lies entirely within the limits of a later invasion and the later deposits have concealed it.

There is, in Western Indiana along the Wabash, a loess of more recent date than the main deposit, but it is confined to low altitudes, seldom appearing more than one hundred feet above the river level. In Western Illinois, a loess has been found which is older than the main deposits, but it has been seen in only a few places and is apparently a thin and patchy deposit. It is thought by Professor Salisbury that the loess of the lower Mississippi was deposited at two distinct stages. Loess is, therefore, a deposit which, like sand or gravel, may be laid down whenever conditions are favorable, but the great bulk of it having been deposited at a definite stage of the glacial period, it seems proper to refer to that stage as the Loess stage.

---
*Eleventh Annual Report, U. S. Geol. Survey, 1889-90, pp. 435-471.

In Southern Indiana, and in bordering portions of Southern Ohio and Southern Illinois, there is a continuous sheet of pale silt locally termed "white clay," which is thought to be a phase of the loess, though more clayey and less uniform in texture than typical loess. It covers the interfluvial tracts as far north as the limits of a later sheet of drift, and has been discovered at a few places beneath that later drift. It probably extended much farther north than its present exposed limits, for the ice-sheet appears to have receded far to the North at the main loess depositing stage, thus leaving the surface free to receive these deposits. The northern limit of the exposed portion in Indiana is marked by the "Wisconsin boundary," shown on the Glacial Map of Indiana. This deposit is usually but a few feet in thickness, seldom exceeding eight feet. Along the Wabash, however, where it becomes a typical loess it often reaches a thickness of twenty to twenty-five feet. It may be readily distinguished from the underlying till both by texture and color. It contains only very minute rock fragments, while the till is thickly set with stones of all sizes. In color it is a paler yellow than the till. There is usually, also, a weathered zone at the top of the till and sometimes a black soil, making still more clear the line of contact.

The loess and its associated silts is found at all altitudes in Southern Indiana; from the low tracts near the Wabash, scarcely 400 feet A. T., up to the most elevated tracts in Southeastern Indiana, which in places exceed 1,000 feet A. T. The great range in altitude is one of the most puzzling features of the loess. The same perplexing distribution is found in Europe as in America. As yet no satisfactory solution for the problem of its deposition at such widely different altitudes has been found.

Interglacial Stage Following the Loess Deposition.—Between the main deposition of loess and the invasion of Northern Indiana by a later ice-sheet, considerable time elapsed; for we find that the drainage lines have reached a much more advanced stage on the loess-covered districts south of the deposits of the later ice-sheets than they have upon those deposits. It is found that large valleys had been opened in the loess and the underlying drift before the streams from the later ice-sheet brought their deposits into the valleys. This interval of valley-erosion is thought by several who have had opportunity to study it, including the present writer, to be longer than the time which has elapsed since the ice-sheet last occupied Northern Indiana.

The question has been raised, whether the greater amount of erosion outside the later drift may not have been due to streams of large volume which accompanied the later ice invasion. That this is only a minor influence, is shown by the fact that valleys in Southern Illinois which lie entirely outside the reach of such waters are much larger than valleys of similar drainage areas within the limits of the later drift-sheet.

It cannot be urged that the region with the smaller valleys is less favored by slopes or stream gradients than the region with well-developed valleys, for the reverse is the case. There are large areas within the loess-covered districts which do not possess the reliefs and other conditions favorable for the rapid development of drainage lines which appear in much of the newer drift. In short, there appears no escape from the view, that the interval between the loess deposition and the later ice invasion was a long one.

The Wisconsin Stage of Glaciation.—After the interglacial interval just mentioned, there occurred one of the most important stages of glaciation in the entire glacial period. It is marked by heavier deposits of drift than those made at any other invasion. Throughout much of its southern boundary in the United States, a prominent ridge of drift is to be seen rising in places to a height of 100 feet or more above the outlying districts on the south, and merging into plains of drift on the north, which are nearly as elevated as its crest.

At this time the ice reached its farthest extension in New England, and also in much of the district between New England and the Scioto river in Ohio. From the Scioto westward, however, it usually fell far short of extending to the limits of the earliest ice invasion. In Illinois it fell short about one hundred miles and in Iowa a still greater distance, but projected into the edge of the Driftless Area in Wisconsin. Partly because of this development in Wisconsin, Professor Chamberlin has called it the Wisconsin Drift Sheet. The limits of this ice invasion appear on the map of North America.

The southern border of this drift-sheet is less conspicuous in Indiana than in the states to the east and west. The ridge on its southern border in Western Indiana rises scarcely twenty feet above the outer border tracts, and it is no more conspicuous in Central Indiana. Indeed, from near Greencastle to the vicinity of Columbus there is not a well defined ridging of drift along the border; the limits there being determined by the concealment of the loess beneath a thin sheet of bouldery drift. From the east border of East White river a few miles below Columbus, northeastward to Whitewater valley at Alpine in Southern Fayette county, there is a sharply defined ridge of drift standing twenty to forty feet above outer border tracts. Upon crossing Whitewater, where the border leads southeastward, it is not so well defined as west of the river, though there is usually a ridge about twenty feet in height.

Although not conspicuous in Indiana by its relief, this border is about as clearly defined as anywhere in the United States. Within the space of a half dozen steps one will pass from loess-covered tracts of earlier drift to the bouldery drift of this later invasion. Accompanying the change from loess to bouldery drift, there is a change in the color of the soil from a pale yellowish or ashy color to a rich black. This line is one of great agricultural importance. The district lying to the north is finely adapted to corn and timothy, while that to the south seems poorly adapted to these crops. The southern district when uncultivated soon becomes thickly covered with briers, a feature which is not common on the black soil of the bouldery drift. In this connection we would remark, that while the loess has usually great fertility, the compact loess of Southeastern Indiana is adapted only to certain products. It seems as well adapted to wheat, orchards, and small fruits as the black soil, and there appears to be an appreciation on the part of the residents of this restricted adaptability.

Between the time when the ice-sheet stood at the line just discussed, and the final disappearance of the ice from Indiana, several moraines were formed. The best defined ones are indicated on the accompanying State map. In a few places not indicated on the map, weak morainic lines have been observed but their courses and connections have not been fully determined.

These moraines indicate considerable complexity of movement, it will be observed that several moraines lead eastward from Illinois into Warren and Benton counties Indiana, and that their eastern ends are crossed by weaker morainic belts carrying many boulders. These features appear to indicate that after the former moraines had been made and the ice had retreated some distance northward, there was a readvance of ice from the northeast to the line marked by the outer boulder belts. It is as yet undecided whether much of an interval of deglaciation preceded this advance, but there was apparently a great shifting of ice-movement.

The prominent moraines which are overridden in Benton and Warren counties may find a continuation eastward in a belt of very thick drift which crosses Central Indiana from Benton county eastward, but which has not the definite ridges which are to be seen from Benton county westward. This belt of thick drift in Indiana is fifteen to thirty miles wide, and has a thickness perhaps three times as great as the general thickness of drift in bordering districts north and south of it. The average thickness is fully 200 feet. It leads south of east across Tippecanoe and Clinton counties to Western Tipton county where it turns abruptly southward through Eastern Boone and Western Hamilton counties and Marion county, coming to White river in the vicinity of Indianapolis. It there turns eastward and passes through Hancock, Henry and Northern Wayne and Southern Randolph counties into Ohio. The belt of thick drift was apparently overridden by the later advance. The weak moraines and boulder belts of the later advance cross it obliquely in a northwest to southeast course in Western Indiana, and return in a northeastward course to it in Henry, Wayne, and Randolph counties.

This later advance apparently extended as far southwest as the bouldery moraine of Central Hendricks county and the bouldery morainic tracts of Southern Johnson and Southern Shelby counties. Its northwest limits were perhaps at the curving belt in Iroquois county, Illinois, and Newton and Jasper counties, Indiana, though there was possibly only a reentrant angle at that line with a Lake Michigan ice-lobe on the northwest.

From this outer limit of the later advance the ice-sheet appears to have shrunk on all sides until its limits on the northwest were at the moraine which lies along the north side of the Wabash in the vicinity of Logansport, and at the southwest were near the dotted line indicated on the Indiana map, leading from White county southeast to the vicinity of Indianapolis. It is in the district southwest of the latter line that feeble moraines and patches of boulders are found crossing over the great belt of drift in oblique courses. From near Indianapolis, the line marking this later position of the last invasion, as shown on the map, leads eastward to the strong belt in Southeastern Delaware county.

There appears to have been at the stage just outlined, a lake bordering the ice on the northwest in which the deposits of sand were made which form such a conspicuous feature in Northwestern Indiana from Cass and White counties northwestward to the moraine north of the Kankakee. It seems probable that the eastern and northern, as well as the southeastern limits of this lake were determined by the ice, for we find that the sandy districts terminate at moraines on these borders.

The moraine leading northward from Northern Fulton county through Western Marshall and St. Joseph counties, would in that case, be about contemporaneous with the moraine on the north side of the Wabash in Southwestern Fulton, Miami, Cass, Carroll and White counties, and both would be of about the same date as the strong moraine lying north of the Kankakee. These correlations are not, however, fully established and should be taken simply as a working hypothesis to be tested by future developments in the study of that region.

In Northeastern Indiana, moraines appear along the north border of the Mississinewa, Salamonie, Wabash and St. Mary's rivers, which were apparently formed in succession as the ice was wasting away after its last advance. These moraines are traceable eastward across Northern Ohio and northeastward into Southeastern Michigan and mark successive limits of a lobe of ice which flowed southwestward across the Erie and Maumee basins. This ice-lobe appears to have persisted at the line of the outer of these four belts to a date when there was open country on the northwest, for the drainage lines lead from this morainic belt northwest to the St. Joseph river, passing across the moraines of the intervening district, as they would scarcely have done had the ice persisted there as long as in the Erie lobe.*

Having traced the ice-sheet to its final disappearance from Indiana, the reader may find in Mr. Taylor's "History of the Great Lakes" † a continuation of the events incident to the retreat of the ice toward Labrador.

SUCCESSION OF ICE INVASIONS SHOWN BY DRIFT DEPOSITS

The evidence of difference in the age of the drift, shown by erosion of its surface, has been discussed. Other lines of evidence of successive invasions have been recognized. One of the most interesting and significant is the presence, in a vertical section, of sheets of drift showing differences of age and of derivation. Such sections are occasionally seen along streams, and are frequently brought to light by wells. Professor Chamberlin has presented as the frontispiece illustration in Geikie's last edition of "The Great Ice Age," such a section found on Stone creek near Williamsport in Warren county, Indiana. There is exposed at the base, a reddish till of the earliest drift upon which there rests a bed of old ferruginous gravel. This gravel is overlain by a fresh blue till, which is apparently of the age of the moraines which lead into that county from the west. Above this till is another gravel bed much fresher than the one below. Above the gravel is a gray till, which was apparently deposited by the ice at the time when it fronted southwest, and had its terminus at the boulder belt which crosses Warren county from north to south just west of the place where this section is exposed.

SUCCESSION OF ICE INVASIONS SHOWN BY STRIÆ

The striæ of Western Indiana, as may be seen by the maps, are widely different in their bearings. Until the several ice invasions had been recognized they were a puzzling feature; but they are now found to support the other lines of evidence of such invasions. Perhaps the best illustration is to be found near Williamsport. There are found in this village two sets of striæ: one bearing southeast and belonging apparently to the earliest invasion; another bearing southward and belonging apparently to the same invasion which formed the bulky moraines in that vicinity. Two miles east of Williamsport, on the north side of the Wabash, Professor Chamberlin found a third set of striæ, with westward bearing, which apparently pertain to the last invasion of the ice.

At Monon and near Kentland, striæ of two distinct sets appear. The latest bear westward and belong, apparently, to the last ice invasion. The date of the earlier, southward-bearing striæ, is as yet undetermined.

THICKNESS OF THE DRIFT

There are surprising differences in the thickness of the drift within

---
* See 18th Report Ind. State Geologist, pp. 29, 89.
† See page 90.

the state. The portion of the older drift exposed to view has, as already noted, an average thickness of about thirty feet. The additional 100 feet of the newer drift is, however, deposited very irregularly. In the belt of thick drift which leads from Benton county southeast to Marion county, and thence east into Ohio, the thickness is probably 200 feet. The portion of the newer drift area to the south of this belt has an average of about fifty to seventy-five feet. A still larger tract extending north from this belt of thick drift as far as Allen county and the west-flowing portion of the Wabash, has only fifty to seventy-five feet with limited areas where its thickness is but twenty to thirty feet. In Northwestern White, Southwestern Pulaski, and Southern Jasper counties there are several townships in which scarcely any drift appears excepting boulders and sandy deposits. In Northern Indiana the drift is very thick. Its average thickness for fifty miles south of the north boundary of the state is probably not less than 250 feet, and may exceed 300 feet. At Kendallville it is 485 feet, and at several cities on the moraine which leads northeast from Fulton county to Steuben county, its thickness has been shown by gas borings to exceed 300 feet. The rock is seldom reached in that region at less than 200 feet. Were the drift to be stripped from the northern portion of Indiana its altitude would be about as low as the surface of Lake Michigan, though much of the present surface is 200 to 300 feet above the lake.

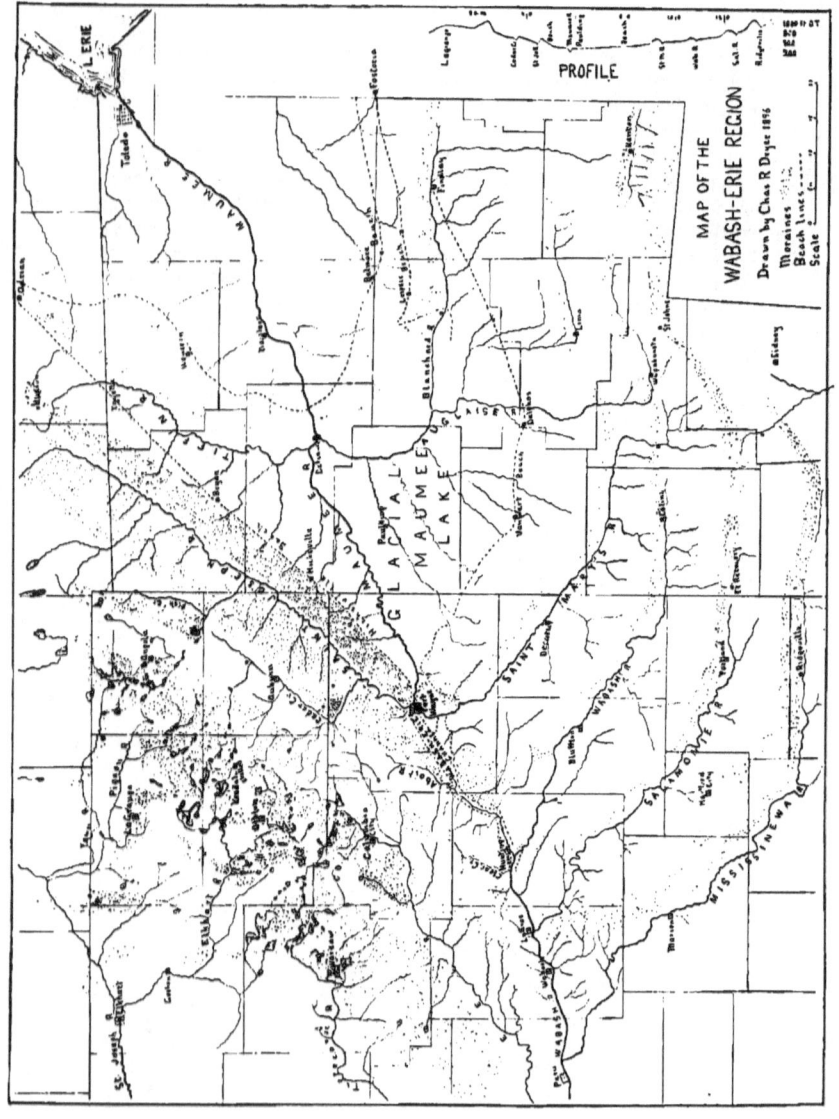

V.—THE ERIE-WABASH REGION

CHARLES R. DRYER

The Erie-Wabash Region is a broad, shallow trough extending from the west end of Lake Erie southwestward across Ohio to Central Indiana. An inspection of the accompanying map will show that it is bounded on the south by the divide between the tributaries of the Ohio and those of the Maumee and Wabash, and on the northwest by a belt of hills forming in part the divide between the Maumee and Wabash drainage and that of Lake Michigan. From Lake Erie the valley bottom rises about 200 feet in 100 miles to a summit near Fort Wayne, and then declines westward, reaching again the level of Lake Erie near Logansport, a distance of sixty miles. The elevation of the southern rim lies mostly between 400 and 500 feet above Lake Erie, while its northern rim rises to an equal height in the hills of Northeastern Indiana. Its width from north to south is a little over 100 miles. This region presents some unique and anomalous features, and exhibits a continuity and uniformity of structure which mark it as an interesting physical unit. Its peculiarities are most clearly revealed by its drainage.

DRAINAGE

An inspection of the map shows that the axis of the trough is traversed by one uninterrupted river channel, occupied, however, by different streams; from Lake Erie to Fort Wayne by the Maumee, thence for about ten miles by a marsh (now drained), thence by the Little Wabash to the main Wabash, and thence by the latter river. Down the sides of the trough flow eight streams of considerable size, four of them arranged opposite each other in pairs—the Blanchard-Auglaize and the Tiffin, the St. Mary's and the St. Joseph. The series on the south is continued at regular intervals by the upper Wabash (above Huntington), the Salamonie and the Mississinewa; but on the north the divide is too near to permit the presence of any large stream except the Eel, which flows more nearly parallel with the axial stream. The drainage system as a whole is almost sagittate, like an unsymmetrical spear-head with five barbs. The general course of these streams is toward the western end of the trough, and normally all ought to be tributaries of the Wabash, yet four turn back upon themselves in a remarkably manner. The St. Mary's, after flowing northwestward sixty miles, and the St. Joseph, after flowing southwestward eighty miles, unite to form the Maumee, which then turns abruptly to the northeast, so that at Fort Wayne the St. Joseph changes its direction more than 160 degrees. The Blanchard-Auglaize flows westward from Findlay about fifty miles, and then by a broad curve nearly reverses its direction. The headwaters of all the southern rivers start directly toward the axis of the trough and Lake Erie, to the northeast, but after a few miles, apparently meet some obstruction and turn at right angles to the northwest. On their right banks tributaries are conspicuous by their absence or brevity. The Maumee from Fort Wayne to Defiance is a sluggish and very meandering stream flowing in a trench thirty to fifty feet deep and without any flood plain. On the south all the streams within thirty miles rise near the St. Mary's and flow parallel with the Maumee into the Auglaize. On the north side of the trough the drainage is almost equally peculiar. All the tributaries of the Tiffin are on the west side, and rise within two or three miles of the St. Joseph. The drainage area of the St. Joseph is, likewise, on its right bank only, and some of its longer tributaries, like Cedar and Fish creeks, have a habit of flowing parallel with it for half their course, and then turning toward it. To join the Aboit or the Eel would seem a more natural course for them. In short, the Erie-Wabash trough does not slope from one end toward the other, as river valleys usually do, but from the middle toward both ends. The valleys of the larger streams have a very long slope on one side, and a very short one on the other: one-half the rivers flow toward, and the other half away from their final destination, and the smaller tributaries, rising near some river which they do not enter, seek a distant outlet by circuitous and troubled courses which make the map a puzzle worthy of Central Africa.

Explanation

It is the business of the scientific geographer to solve all such puzzles, and not to rest until the cause and origin of every phenomenon is explained. He is not content with discovering, describing and mapping natural or artificial features, but proceeds to ask and answer, if possible, the question, How did these things come to be so? An expert geographer could now infer the principal features of relief and the main events in the history of the region from a map showing the streams only. It is easy to see that along the Erie side of each of the large tributaries of the Maumee and the Wabash a ridge of some kind must exist, forming a barrier and a divide. It is now much easier to guess the nature of these ridges than it was in 1870, when Mr. G. K. Gilbert, then a young geologist employed upon the Ohio survey, announced that he conceived the ridge along the Erie side of the St. Joseph and St. Mary's rivers to be "the superficial representation of a terminal glacial moraine, that rests directly on the rockbed, and is covered by a heavy sheet of Erie clay—a subsequent aqueous and iceberg deposit." This was the key to explain all the peculiarities of the region, and subsequent observers have had little more to do than to apply it. Since 1870 the whole complex morainic system which extends across the United States from Cape Cod to Dakota has been surveyed and mapped, and the relation of the ridges of the Erie-Wabash region to that system has been determined.

Moraines

A glance at the map will show the peculiar form and arrangement of the moraines. They are seen to constitute a quite regular and symmetrical series of crescentic ridges, parallel, in the main, with each other and with the southwest shores of Lake Erie. Upon the north side of the axis they are crowded together, straightened out and otherwise deformed.

The Blanchard-Tiffin Moraine, nearest to Lake Erie, and the youngest of the series, probably never very massive, has been so modified by the action of lake waters as to be the least prominent of all, but an expert glacialist would have little difficulty in tracing its course across the otherwise level country from Adrain, Michigan to a point east of Cleveland, Ohio. It is more conspicuous east of Findlay than west of that place, and in that stretch consists, according to Mr. Leverett, of "a broadly ridged and slightly undulating tract of till (stony clay) standing twenty to forty feet or more above the plain south of it, and having a breadth of one and one-half to three miles." From the point where it is crossed by the Leipsic beach to the Maumee the moraine has a comparatively smooth surface. North of the Maumee it occasionally spreads out into a broad tract of sand. It rises and falls in its course across the country paying little attention to levels, but varying between 730 feet above tide near Defiance to 1250 feet at its eastern end.

The St. Mary's-St. Joseph Moraine is more massive and uniform than the one just described, yet it seldom presents any striking features which would attract the attention of the passing traveler. It is best seen upon the map by its influence upon the course of streams. "It is like a dead wave upon the surface of the ocean, hardly perceptible to the eye on account of its smoothness, but revealed by its effect upon everything that encounters it." It is not, as Mr. Gilbert thought, a buried moraine, but the ridge he describes *is* the moraine from bottom to top. South of the Maumee its Erie slope is very gentle and merges imperceptibly into the plain. Its crest is fifty to eighty feet above the St. Mary's river to which it slopes more abruptly. Its surface is generally smooth but occasionally becomes rolling with bluffy margins. It is composed almost entirely of glacial clay, sand and gravel being very scarce. North of the Maumee the moraine is direct in its course, and its margins are sharply defined by the beach line on one side and the St. Joseph river on the other. It is four or five miles wide, fifty to seventy feet high and of the same structure as the southern portion. To the north, like all the rest, it connects with the moraine system in "the thumb" of Michigan. Its slope lengthwise is quite regular, from 800 feet above tide at Ft. Wayne to 900 feet in Southern Michigan and at its eastern end in Ohio.

The *Wabash-Aboit Moraine* can be traced continuously a greater distance than any of its neighbors, at least from Central Ohio westward into Indiana, and northwestward far into Southern Michigan. In general character it resembles the St. Mary's-St. Joseph moraine. The upper Wabash once followed its outer face from Celina to the axial channel, but now turns away from it below Bluffton toward Huntington. The northern wing is much broader and more massive than the southern, and fills the space between the St. Joseph river and a half-filled valley partly occupied by the Aboit river, Cedar creek and the headwaters of Pigeon river. It is a rolling table-land, five to fifteen miles wide, about 100 feet above the St. Joseph and fifty feet above the valley on the west. Occupying the middle position in the series, it assumes some of the characteristic features of its neighbors on the west. Although its chief material is clay, it has a habit of rising here and there into abrupt rounded hills or conical peaks of gravel fifty to one hundred feet above the general level, as at the elbow of Cedar creek and in Southeastern Steuben county. These hilly portions are accompanied by small lakes, which become more numerous as the Michigan line is approached and passed. Fish and Clear lakes, both in Steuben county, and among the largest and most beautiful in Indiana, belong to this moraine. Like the rest, it rises from its apex, 870 feet above tide in Southwestern Allen county to 1100 feet in Michigan and 940 feet in Hardin county, Ohio.

The *Salamonie-Blue Moraine* is easily traced upon the map from "the knot" near Kenton, Ohio, along the usual curve to Angola, Indiana, but compared with the others it is weak, diffused, and inconstant. In Huntington county it is broken up into several strands and in Southern Whitley is represented by a belt of boulders. North of Eel river it can be distinguished from the general morainic mass only by its small features. It is a tumbled country; the hills, hollows, mounds, saucers and lakes are all there in great number and variety, but in miniature. It contains however one lake of the larger class—Blue River lake in Northeastern Whitley. Its elevation varies from 800 feet in Huntington county to 1050 feet at Angola, and 1060 at St. Johns, Ohio.

The *Mississinewa-Eel Moraine* surpasses the others in symmetry, and curving from "the knot" far southward, sweeps through a semicircle of full 200 miles. If one wishes to see a terminal moraine in all its distinguishing peculiarities, to get intimately acquainted with all its moods and phases so that he would be able to recognize one if he came across it anywhere, he cannot do better than to walk or drive over Whitley, Noble, Lagrange and Steuben counties, Indiana. The typical portion is from ten to twenty-five miles wide and sixty miles long. It is an irregular, variously undulating pile of clay, sand, gravel and boulders with the coarser materials predominating. Its surface is 150 to 300 feet above the country on either side, and its total thickness down to bedrock from 200 to 475 feet. Its topography defies verbal description in detail, but may be included under a few general types. The greater part of the area may be designated as *crumpled*, resembling the surface of a sheet of paper which has been carelessly crushed in the hand and then spread out. The ridges have no particular direction, their tops are broad and slopes gentle, yet there is very little level ground. This type passes by insensible gradations into the *corrugated*, in which the ridges are steeper, sharper and arranged in somewhat parallel lines. Similar features very much exaggerated produce what may be called *gouged* or *chasmed* country, found in perfection southwest of Columbia City. The surface is entirely occupied by deep, irregular, elongated valleys, with narrow, sharp, winding ridges between, all in indescribable confusion. The roads through it are very crooked in order to avoid the marshes, yet, in every direction, they are a series of steep descents and ascents. The relief might be imitated by taking a block of plastic clay and gouging it with some blunt instrument in as irregular a manner as possible.

Scarcely more extreme and peculiar is the topography usually regarded as typical of terminal moraines, "the knob and basin." It consists of confused groups of dome-shaped or conical hills, often as steep and sharp as the materials, usually sand and gravel, will lie, with hollows of corresponding shape between. The impression made is as if the material had been dumped from above and left as it fell, like gravel from a wagon. Some of the finest specimens in America occur south of Albion, in the Diamond Lake hills near Ligonier, east of Lagrange, in the northwest corner of Lagrange county, and the grandest of all, north of Angola, where the peaks rise to about 1200 feet. Throughout this morainic region the hollows or "kettle-holes" are occupied by marshes or lakes, the largest of which are shown upon the map. The number of such lakes in Indiana must be more than a thousand, and the marshes, or extinct lakes, out-number the living ones. A description of these will be given in another chapter. The last moraine, like the rest, rises towards its extremities, from 700 feet at La Gro to 1200 feet in Steuben county and 1100 feet in Ohio.

The peculiar structure of the Erie-Wabash region may be summed up in the following statements: Between Lake Erie and Peru the Erie-Wabash trough is crossed by five morainic ridges, which sweep in wide curves from one side to the other, parallel with each other and with the western shores of the lake. They vary in form from sagittate to crescentic with their convexities to the southwest. Their apexes rise with the general slope of the trough from the first to the third, and fall from the third to the fifth, while the extremities of the wings rise successively to greater heights through the whole series. The northern wings are more massive and are crowded together, while the southern are more symmetrical and disposed at equal intervals except that between the first and second which is greater than the other intervals. The southern tributaries of the Wabash and Maumee flow along the outer faces of these moraines, the northern more irregularly follow the narrow intervals between them. The profile upon the right side of the map, drawn from Lagrange to Paulding and from Paulding to Ridgeville in order to cross northern and southern moraines and streams at right angles, shows these peculiarities in a striking manner. Thus is the strange behavior of the streams which has puzzled observers for a hundred years explained, but only by substituting another puzzle in its place—that of the moraines. They remain to be accounted for, a problem which, like many others in geography, can be solved only by reading backwards into the remote history of the region.

Physical History

The studies of hundreds of geologists during the last twenty-five years have established the fact that a large part of North America was once covered by an ice-sheet which moved from its gathering grounds around Hudson Bay southwards to the Ohio and Missouri rivers.* The *motion* of the ice was always forward to the South, and its retreat was accomplished by the melting back of the front edge; thus, even while the edge was retreating the ice itself was always moving forward. Sometimes it came faster than it melted, sometimes the supply just balanced the melting, sometimes it melted faster than it came. The ice was thicker in the old valleys than upon the divides, and the valleys offered an easier course. Consequently it advanced farther in the valleys and retreated from them more slowly, and the sheet came to have a lobed or scalloped edge. The Erie-Maumee-Wabash valley contained one of these lobes which extended far into the heart of Indiana.† As the growing warmth of the climate melted it and the diminishing volume of the ice-stream from the North became insufficient to supply the waste, the edge slowly retreated. The whole load of soil and stones which it had gathered on its way from Labrador was left evenly spread out over the old rock surface, as a sheet of drift or *ground moraine*. But its retreat was not continuous. For some unknown reason there were periods when the supply of the ever-advancing stream

*See map, p. 33.
†See map, p. 28.

was equal to the melting, and during such periods the edge remained stationary along a certain line. In that case the melting ice dropped more of its perpetually arriving load at the edge than elsewhere, and thus built up a ridge or *terminal moraine*. Such a morainic ridge, then, marks the line at which the edge of the ice-sheet halted for a long period, and thus reveals to us the shape of the lobe. In the Erie-Wabash region a succession of halts and retreats was performed with great regularity. The movements were like those of an army retreating in good order, which alternately throws up breastworks along its line of battle and abandons them to fortify another line farther back. To vary the simile a little, it is like an army ever advancing in solid column, but on account of the hot fire of the enemy not a man ever gets beyond a certain line, and at intervals the head of the column for a long distance back is wiped out. In such a case the distribution of bones, weapons and accoutrements would be precisely like that of the glacial drift.

The story of events on the north side of the Erie ice lobe is a little different. There it ran against the side of another ice lobe which moved from Saginaw Bay southwestward into Northern Indiana. Its advance was obstructed, the ice was piled up in a thicker mass and the retreat was slower. The results of this are evident in the greater massiveness, straightness and crowding together of the moraines. The outermost moraine is the joint product of the Erie and Saginaw lobes, which accounts for its strongly marked features as before described. The line between Erie and Saginaw drift can be approximately located as passing through Albion, north and northeast through the corner of the four counties to the western border of the moraine in Northern Steuben. There is a notable difference of topography, soil, forest and flora upon the two sides of it. The moraines shown upon the map west of this line belong to the Saginaw lobe, which was comparatively feeble and disappeared from the state earlier than the Erie. The glacial invasion of Northeastern Indiana is a story of advance in double but unequal columns; of prolonged struggle between them; of defeat and evacuation on the part of the weaker forces, and of deliberate retreat on the part of the stronger from the field of battle.

GLACIAL DRAINAGE

The melting ice, of course, furnished a large supply of water, which in its escape, established not only the present drainage channels but many others now abandoned. The Eel River moraine contains scores of glacial drainage channels now partly filled with drift, and occupied by a lake, marsh or small stream. None of the present stream channels are older than the period of melting ice. The present streams frequently fail to fit the channels made by the original swollen floods, and wander about alternately in valleys much too wide for them, and in narrow ravines of their own cutting. The evidence is abundant that the Saginaw ice got out of the way first and left the country open for free drainage from the Erie ice into the basins of the St. Joseph of Lake Michigan and the Kankakee. The Mississinewa and Eel rivers were born at the same time, but after the first step of retreat the latter was lengthened by the addition of the Blue river and kept its channel open through the moraine at South Whitley. At this stage the Salamonie began. After the next retreat the Upper Wabash carried the drainage of the ice front, assisted on the north by the Upper Pigeon river, possibly Fish creek and Cedar creek; the latter probably joined the Aboit, since the Aboit Valley is far too large to be the work of the present short stream alone. When the ice retreated to the St. Mary's-St. Joseph moraine and the corresponding rivers came into existence, the Wabash received a considerable accession to its length which placed its source in Southern Michigan. At this period the waters of the two rivers were carried southwestward through the Erie-Wabash channel, or, perhaps it is nearer the truth to say that these waters then cut that channel.

Lake Maumee

As soon as the ice-front began to retreat from this moraine and to uncover the present Maumee Valley, the slope was *toward* the ice, and the water began to stand between the moraine and the ice-front in the form of a long, narrow lake. Although this lake found an outlet westward into the St. Joseph-Wabash, it grew in area and depth until an adjustment between the level of the outlet and the ice-wall of the glacier while it paused at the line of the Blanchard-Tiffin moraine, caused the lake to occupy a relatively permanent position. This was Lake Maumee, as odd in shape as in other conditions. It had for its eastern shore a wall of ice extending from Findlay through Defiance to Adrian, for its southern shore a beach of its own construction now called the Van Wert ridge, and for its northwestern shore the margin of the St. Joseph moraine, along which its waves built up a beach now known as the Hicksville ridge. It emptied its surplus waters through the Ft. Wayne outlet and Erie-Wabash channel into the present Wabash at Huntington, forming a river a mile wide and deep enough to rival the present Niagara. This lake must have existed for many years or centuries, but could not be permanent; for the dam which held it up to that level was of ice. The ice continued to melt and retreat until an outlet was uncovered at a lower level in "the thumb" of Michigan, and the water began to flow that way into Saginaw Bay. The lake level slowly fell, the Erie-Wabash river began to dwindle, its channel silted up, and finally, after the lake was completely drained, even the St. Joseph and St. Mary's turned back through the gap and the Maumee river was born. A great thickness of fine clay had been deposited in the bottom of the glacial lake, through which the Maumee and other streams had to cut their way. The surface was very level, and the sluggish streams had to wriggle over it in tortuous courses. To this day the old lake bottom is one of the most difficult areas in America to drain.

Surface and Soil

We are now prepared to account for the variety of surface and soil found in the Erie-Wabash region. Underneath it all is the *ground moraine* of rather stiff gravelly clay, similar to the general mass of the drift sheet, and forming the surface soil over the greater part of the area. Piled upon this are the *terminal moraines*, largely composed of the same materials but containing, locally, great heaps and masses of sand and gravel. Out of this the glacial and present streams have washed much of the finer material and deposited it in the old valleys and intermorainic intervals, some of which are nearly filled up; as, along the upper Pigeon river, Cedar creek and the head of Eel river. Innumerable hollows and depressions which at first contained shallow lakes, have been filled with vegetable growth and converted into marshes, or if sufficiently drained, *muck lands*, rich for grass and corn. The bottom of the glacial Maumee lake is an exceedingly fine tough clay to which, in many places, the growth and decay of vegetation have added improved qualities.

Culture

The whole region was originally covered with a heavy growth of hardwood forest, except the marshes, or so-called "wet prairies," and a few small tracts of genuine dry prairie in the northwest. No equal area has furnished more valuable timber, oak, walnut, beech, maple, ash, elm, sycamore, poplar, hickory, locust, cherry and others. For unknown centuries before the advent of the white man, the Indian hunted in the forests and fished in the lakes. The Maumee-Wabash was an important route of canoe travel between the Great Lakes and the Ohio. The carry or portage from the head of the Maumee over to the little stream which now occupies the Erie-Wabash channel, was short and easy, and in 1680 LaSalle found there an Indian village and a fur-trading post. Here was a favorite congregating place for men, savage and civilized, at the forks of four water-ways, and the spot was naturally predestined to be the site of an important town. It has passed through all the regular stages characteristic of so many American towns, Indian village and portage, trading post, military fort, modern city. It was as easy a route for the canal boat as for the canoe, and as early as 1834 the Wabash and Erie canal was constructed through it, having its summit level in the

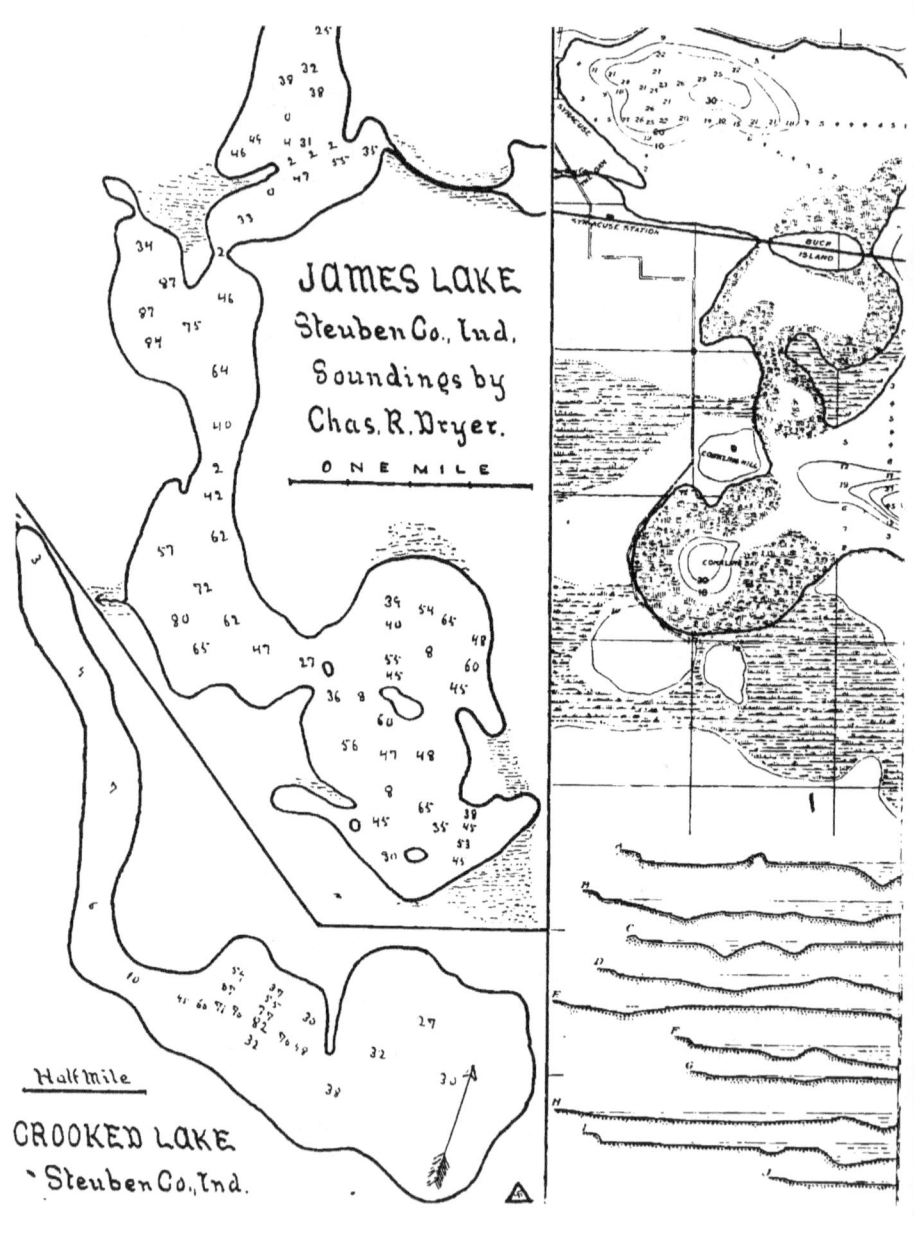

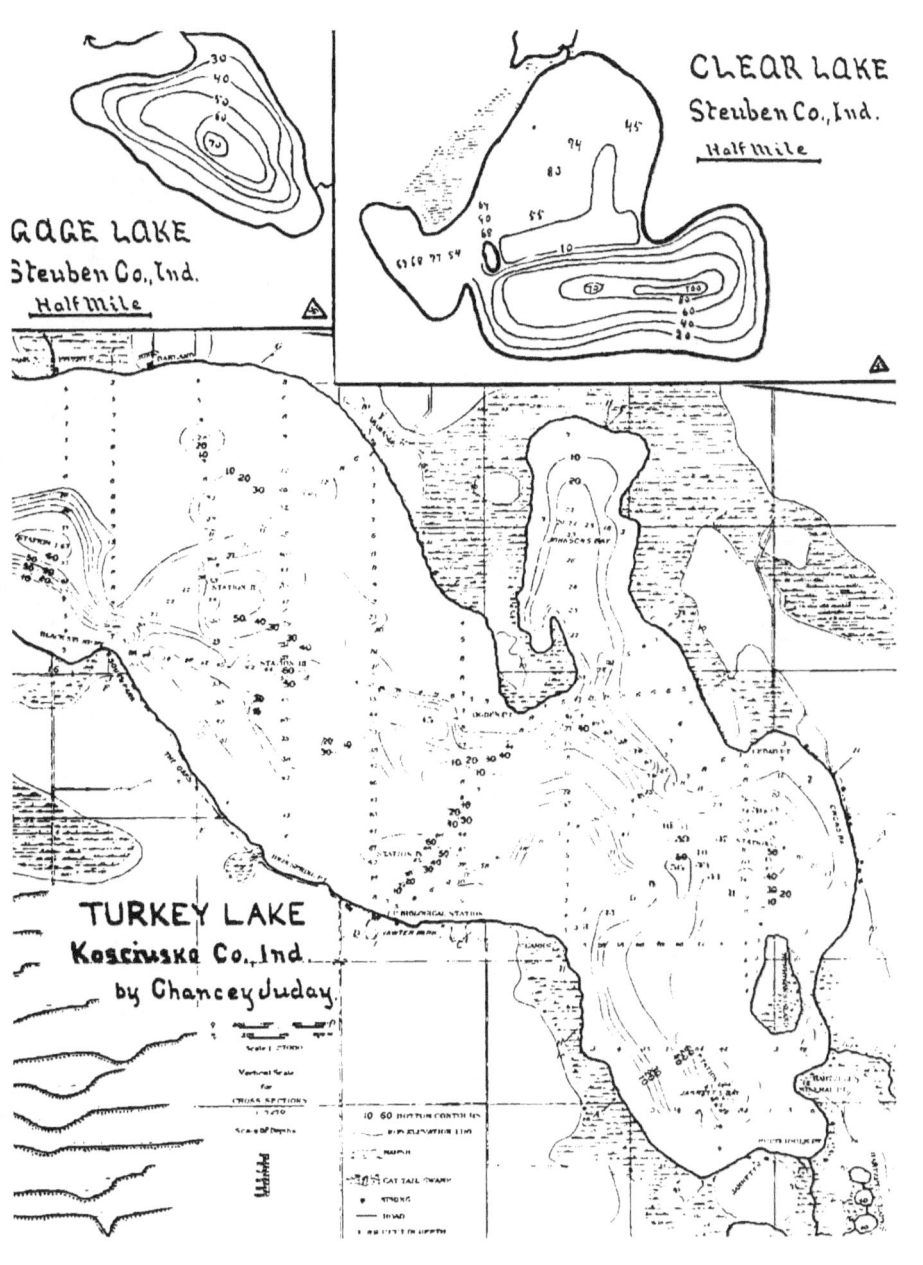

VI.—THE MORAINIC LAKES OF INDIANA

CHARLES R. DRYER

An intelligent young man once told the writer that he had taught school in Indiana ten years without knowing that there was a lake in the state, yet his pupils probably learned something about Titicaca and Tanganyika. This is a not unusual case of the prevalent love of the remote which afflicts the teaching of geography. That there may be fewer such teachers and pupils in the future, is one of the objects of this paper.

DISTRIBUTION

Nearly every map of Indiana shows some of the lakes but none gives an adequate idea of their number. They are most numerous in two belts; one extending from Steuben county to Fulton, the other from St. Joseph to Lake. An examination of the glacial map of Indiana* shows that these lake belts coincide with the great interlobate moraines formed between the Michigan, Saginaw and Erie ice lobes. There are very few lakes outside the area of influence of the Saginaw ice. The Indiana lakes are a part of the great morainic lake belt which extends from Cape Cod to Dakota, and in no portion of that belt are the lakes more numerous and characteristic. Steuben county, alone, contains more than one hundred, and the whole number in the state cannot be less than one thousand.

CLASSIFICATION

Glacial lakes are of two classes: (1) *rock basins*, formed wholly or partially by glacial erosion; (2) *drift basins*, formed by the irregular deposit of drift. The former are very numerous in Canada, New England, Scotland, Sweden, Finland and regions of ice accumulation generally. The latter are characteristic of regions of ice destruction and drift deposition, as the North Central States, and North Germany and Russia. No glacial rock basin occurs in Indiana, and under the most of our lakes the drift is probably not less than one hundred feet deep. They all belong to the class which Davis† has called lakes of *obstruction*, as distinguished from basins formed by construction or destruction. In general, they may be said to be due to the irregular deposit of glacial drift; the hollows or

---
*See page 28.
†Proceedings Boston Society of Natural History, Vol. XXI., p. 315.

basins being the counterparts and complements of the hills and knobs characteristic of terminal moraines.

Penck* divides morainic lakes into three classes: (1) Round, cauldron-shaped basins, known in this country as *kettle-holes*, or "potash kettles," many of which are dry. (2) Long, narrow channels containing shallows and deeps like the beds of rivers, which they evidently once were. (3) Basins which are branched, lobed or otherwise irregular, often extremely so, and whose bottom topography is undulating like the surface of the land around them. To these might be added basins of complex origin which combine some of the characters of the three classes.

SIZE

Morainic lakes are always small, the area of the majority being less than one-fourth of a square mile. The largest in Indiana has an area of a little over five and one-half square miles, while the Mauersee, in East Prussia has an area of thirty-five square miles, divided, however, into six basins, and a maximum depth of 125 feet. The depth varies from a few feet to a little over 100 feet, which, in some small lakes, makes the slope about as steep as the material will lie.

KETTLE-HOLE LAKES

One of the finest specimens of a lake with a single, symmetrical, kettle-shaped basin is Gage lake in Mill Grove township, Steuben county. (See map.) It is about one mile by three quarters in diameter, and surrounded by high sand bluffs. The slope of the bottom is quite uniform from every side, and a large area in the center is over fifty feet in depth, with a maximum of seventy feet. Clear or Pretty lake, in Milford township, Lagrange county, is about the same size as Gage, and its basin, nearly circular in outline, forms a perfect washbowl eighty feet deep in the center, gradually shallowing to about sixty feet towards the shore in all directions, then rising rapidly to a wide, shallow rim all around. Blue River lake, in Smith township, Whitley county, belongs to the same class but is larger and less deep.

Clear lake in Clear Lake township, Steuben county, is a double, or perhaps triple kettle-hole, divided by a ridge which rises to six feet below the surface. (See map.) Its area is 1.18 square miles. The south basin is regular in outline, a mile and a quarter long by half a mile wide. There is a coast shelf of shallow water, from which the bottom falls away rapidly, the slope being in several places as much as one foot in two, or at an angle of more than twenty degrees. At one place the depth increases in ten boat-lengths from six, to ninety feet. A large portion of the central area is below sixty feet, and the line of greatest depth

*Morphologie der Erdoberfläche, II., 265.

varies between seventy and one hundred feet. The water is very clear, and reported by divers to be very cold in some places at the bottom. Over these areas ice seldom forms, and they probably indicate the position of copious sub-lacustrine springs. To this class also belongs some members of an interesting group of lakes in Johnson township, Lagrange county. Two of the group, Atwood and Witmer, are situated within a terminal moraine of the Saginaw glacier, and are surrounded by high hills, but are quite shallow. The others are in a level intermorainic interval. Third lake is an irregular hole of perhaps 300 acres in the midst of an extensive marsh. A depth of ninety-six feet was found within twenty rods of the inlet, and no water beyond was found less than seventy-five feet deep. Oliver and Olin lakes, about 600 acres in area, lie in the same level interval, but not in a marsh. As far as sounded, they proved to have a quite uniform depth of from sixty to eighty feet. These deep, abrupt and smooth-bottomed basins, not among the hills, but sunk into the level surface of the ground moraine, upset the supposed rule that lakes with low shores are shallow.

Examples of kettle-hole lakes might be cited indefinitely. They are of all sizes, from a mere pool up to one or two square miles. Dry kettle-holes far outnumber the lakes, and are of all dimensions, from a mere dimple, saucer or soap-dish to a great cauldron or funnel. The writer has seen in Western New York, near the summit of a morainic gravel hill, a perfect funnel about two acres in area at the top and tapering downwards 100 feet to a sharp point. On account of the porous nature of the soil it never retains, even temporarily, a pool of water. If a kettle-hole sinks into the clayey ground moraine or is lined with an impermeable clay deposit, as a cistern is lined with mortar, it will usually be filled with water up to the level of the lowest point in its rim, and if the rain-fall exceeds evaporation, will have an outlet. If it sinks into sand or gravel below the level of permanent ground water, it is like a well, and will hold water up to that level, but will not overflow. The celebrated and marvelously beautiful Walden Pond in Concord, Massachusetts, rendered famous by Thoreau and Emerson, is a kettle-hole lake in a glacial sand plain, sixty-five acres in area, 100 feet deep, and without visible inlet or outlet.

Origin.—The precise mode of formation of kettle-holes was for a long time a puzzle, until observations of existing glaciers revealed the process. During a period of glacial retreat the ice near the margin is stagnant and covered with debris to a considerable depth. Large masses of ice become detached from the main mass and, buried in drift, are left to melt. As they slowly disappear, the drift material caves in over the vacant space and only a hole remains, its depth, dimensions and slope depending upon the thickness and breadth of the ice block, and the character and quantity of the moraine material. Kettle-holes, both dry and water-holding, are among the most characteristic and easily recognized features of terminal moraines.

CHANNEL LAKES

Terminal moraines contain many long, narrow lakes, which occupy valleys generally much too large for them, and have uneven bottoms with alternating deeps and shallows like old river-beds. During the recession of a glacier large volumes of water flow away from the ice front and carve deep channels for themselves in the loose moraine material. After the disappearance of the ice these channels are abandoned, or, being supplied only by rainfall, the volume of the stream is greatly diminished. They partially fill up with sediment, and come to be occupied by marshes or shallow lakes, threaded and connected by an insignificant stream. As has been elsewhere noted,* the Saginaw ice lobe withdrew from Indiana while the Erie lobe still occupied the northeastern portion in considerable strength; and the whole northwestern slope of the joint interlobate moraine in Steuben and Noble counties is furrowed with glacial drainage channels. In Steuben county several transverse valleys cut entirely through this moraine and carry water from the interval on the Erie side into the Lake Michigan basin. They are a half mile to a mile in width and 150 feet deep, and each contains a chain of lakes strung upon the thread of a small stream.

The larger lakes of these chains are mostly of complex structure and origin, but many of them are typical channel lakes. The long, shallow arm or neck of Crooked lake in Pleasant township (see map) is a perfect example of this kind. Long and Golden lakes of the Pigeon river chain in Steuben township are each more than a mile long and scarcely one-fourth of a mile wide, with a middle depth varying from twenty-five to forty feet. From Hogback lake, the next below in this chain, a similar valley trends northward five miles to Gage lake of the Concord creek chain. This, too, was once an important drainage line, but a number of sand and gravel ridges a few rods wide and thirty feet high, resembling a railroad embankment or fill, have been in some way thrown across the valley, and ponded between them are half a dozen shallow pools without outlet. A similar phenomenon is presented by the valley of Long lakes in York township, Noble county. Long lake in Milford township, Lagrange county, two miles long and nearly half a mile wide, probably belongs to this class, but is of unusual depth—forty-five to eighty feet. Shriner's and Cedar lakes in Thorn Creek township, Whitley county, occupy two narrow, parallel valleys, separated by a ridge scarcely a quarter of a mile wide. Shriner's is straight and symmetrical, one mile by one-fourth, its middle depth increasing from forty feet at the

*See page 19.

foot to over seventy near the head. Cedar is much more irregular in outline and bottom, and is divided by shallows into two basins, of which the upper is nearly eighty feet deep. Round lake, 100 acres in area, and 60 feet deep, connected with Cedar by a narrow channel and at the same water level, is probably a kettle-hole.

Irregular Lakes

Lakes of lobed, irregular and complex form and outline, are numerous. They may have been formed simply by the irregular, tumultuous dumping or heaping up of drift, but many are probably of complex origin, including within one connected area kettle-holes, old river channels and basins due neither to the melting of detached ice blocks, nor to stream erosion. No better example exists in the world than James lake, in Pleasant and Jamestown townships, Steuben county. (See map.) It consists of five distinct basins, with a total length of five miles and an area of 2.21 square miles. The southern and largest basin is one mile by a mile and a quarter, with very irregular shores and bottom. Three small islands stud its surface, and at another point a mound in the bottom rises to within eight feet of the surface. The depths between vary from thirty to sixty-five feet. Upon the east side the shores are abrupt, and the hills rise steeply to a height of one hundred to two hundred feet. Bold promontories, sequestered coves and precipitous bluffs give it a highly picturesque character. The second basin is more regular, with a length of one mile and a maximum width of half a mile. The east shore continues to be high and steep, and only a few rods from it sixty feet of water can be found. The maximum depth is eighty feet. Northward it narrows to a strait with only two feet of water, opening into the third basin, which in shape, size and depth, closely resembles the second basin. Eagle island, a high peak rising abruptly from the water near the north end, is now joined to the mainland by a marsh. A few rods off its west shore the deepest sounding in the lake was made, eighty-seven feet. A narrow passage leads to the fourth basin, which is continuous to the east with a valley, which cuts completely through the moraine and contains numerous small lakes, surrounded by extensive marshes. Its depth varies from thirty to fifty-five feet. This basin is bounded on the north by Deer island, similar to Eagle, and a bar thickly overgrown with rushes. The lake seems to end here, but if one pushes through the rushes, he emerges into the fifth basin, larger than the fourth and about the same depth. The valley continues northward several miles into Michigan, and contains Lake George, as large as the southern basin of James, besides many small pools. These are drained by Crooked creek, which again emerges from James lake on the west side of the second basin, and in less than half a mile empties into Jimerson lake. The whole connected series of basins seems to occupy three valleys, which were important lines of glacial drainage: one from the southeast through the first and second basins of James and Jimerson, one from the north through George, and the fifth, fourth and third basins of James, and one from the east into the fourth basin. The space between the east and southeast valleys, occupied on the map by the label, contains the highest, most precipitous and irregular group of morainic knobs in Indiana—rising at one point to 1,200 feet above tide. The level of James lake is about 1,000 feet. The whole region is as nearly Alpine in character as moraine topography can be, and though Alpine only in miniature, it presents a surprising variety of scenery, which rivals many more famous localities.

Among the morainic lakes of Indiana, James lake is surpassed in size only by Turkey lake in Kosciusco county, which has recently been thoroughly surveyed by Messrs. Juday and Ridgley, of the Indiana University Biological Station. A report of their survey appears in the Proceedings of the Indiana Academy of Science for 1895, to which we are indebted for the map herewith reproduced, and for the following interesting data. The map tells its own story better than words can. Turkey lake is made up of two parts connected by a channel three-quarters of a mile long and from one to five feet deep. The part north of the channel, known as Syracuse lake, includes an area of three-quarters of a square mile; has an average depth of thirteen and a half feet and a maximum of thirty feet. The greatest length of the main lake is about four miles and its greatest width one and a half miles. The entire shore line is between twenty and twenty-one miles in length, and the area a little more than five and a half square miles. The average depth is computed to be between seventeen and twenty-two feet; the greatest depth is sixty-nine feet. An examination of the contour lines of the map shows that very much of it, an area computed to be three and a quarter square miles, is less than ten feet deep. If the level of the lake were lowered thirty feet the area would be reduced to one and fifteen-hundredths square miles, and it would consist of four bodies of water connected by channels from 100 to 200 feet wide and less than ten feet deep. These would be: (1) A small area in Crow's Bay with a maximum depth of nineteen feet; (2) about one-half of Jarrett's Bay with a maximum depth of thirty-eight feet; (3) the main body of the lake, its width decreased almost one-half, and its maximum depth being thirty-six feet; (4) A small area toward the west end with a maximum depth of thirty-three feet. Lower the level of the lake forty feet and these four bodies of water would be separate lakes. "The similarity of the lake bottom to the surrounding country," remarks Professor Eigenmann, "which seems to have been little changed by erosion, makes it quite certain that the lake basin is due to the irregular dumping in a terminal moraine, parts of it containing deeper kettle-holes." Many interesting data in regard to shores, beaches, outflow, evaporation, temperature, ice, etc., may be found in the report of Mr. D. C. Ridgley before cited.

Life History

Of all the varied features now presented upon the face of the earth there are probably none whose essential characteristics are more obvious, whose life histories are more easy to read than those of the morainic lakes. They are all geologically young, those of Indiana being confined to the very latest moraines of the glacial period. They are mere babes born yesterday, and destined to die to-morrow. During the period of glacial melting it seems certain that all existing valleys, except drainage lines of rather steep slope, would tend to be filled up. At any rate, many such half-filled valleys now exist, and it is probable that all the kettle-holes and basins have suffered a considerable diminution in depth. As soon as the surface became subject only to the wash of rainfall and was covered with forest, general erosion and removal of material from the slopes into the hollows was greatly diminished, and at present the results of these processes are practically nothing. The streams which now empty into the lakes are few and small, and the quantity of sediment thus brought in is very trifling. A recognizable delta is almost unknown. Many of the lakes are great springs fed by inflows at the bottom, and the evaporation so nearly balances the supply that the outlets are small and feeble. Natural down-cutting of outlets is nowhere perceptible. The deposit of lime and iron salts from the overcharged ground-water is probably doing more to fill up the lakes than surface erosion. This phenomenon is more noticeable in some lakes than in others. Aquatic plants are, as a rule, incrusted with lime, and mussel shells and pebbles upon the bottom form nuclei for similar deposits which soon render their original form scarcely recognizable. The water of some shallow lakes seems of milky whiteness on account of the deposit of marl on the bottom, and such lakes look, at a distance, like silver coins or platters laid down among the hills.

Another very efficient agent tending toward the extinction of these lakes is man himself. In the case of small and shallow lakes, artificial drainage has often resulted in their complete destruction, while the areas of large shallow ones have been reduced one-half or more.

A third agency more effective than all others for the obliteration of morainic lakes is the growth of aquatic vegetation. The character and extent of this growth depends somewhat upon the depth of the lake and the slope of the shores, but chiefly upon the nature of the bottom. In this respect lakes may be divided into three classes—lime lakes, sand lakes and peat lakes. In lime lakes the bottom is composed of marl, and all vegetation is very scanty and stunted. This is true to nearly the same degree of lakes with sandy bottoms. But a large majority of the lake-beds are covered with a black, tenacious mud which furnishes the soil for a luxuriant growth of aquatic plants wherever the requisite shallowness and stillness of water permit. Small lakes are often surrounded by a border of dense vegetation which extends out as far as the line of about twelve feet in depth. In the large lakes this occurs only upon the west side, even when the conditions of soil and depth appear equally favorable upon the east side. This is due to the prevailing westerly winds, which create too much wave disturbance along east shores for the accumulation of peat. The lakes are literally being filled with solidified air, the great bulk of the solid material which composes the plants being absorbed from the gaseous ocean above and consigned to the watery depths below. The maps of Steuben county show in Fremont township Cedar lake as being a mile in diameter. In fact, there is no lake there. Some of the water has been drawn off by artificial drainage, and the remainder is now covered by a floating, quaking bog with a few open lagoons. This lake has been buried alive by a growth of peat, and that there are many such in Indiana, the railroad companies which have tried to lay a track across them have found to their cost. Extinct lakes are more numerous than living ones, and their beds are marked by bogs or meadows underlain by fifteen or twenty feet of muck. The process is slow if measured by the years of a man's life; perhaps the peat bed extends into the lake only a few feet in a century. The present dominant race of men may pass away and leave these lakes still lying like bright jewels among the hills; but every one is doomed to final extinction.

> "The hills are shadows and they flow
> From form to form, and nothing stands:
> They melt like mist, the solid lands,
> Like clouds they shape themselves and go."

But of all features of the landscape, lakes are the most ephemeral. As long as they remain they will continue to contribute to the service and delight of man. They fed the savage with fish, but they feed the more highly developed man with beauty, and afford means for that relaxation and healthful pleasure which the conditions of modern life demand. The time may come when the lakes of Northern Indiana will be the most valuable property of the region, and means will be sought for preserving, instead of destroying them. Between the Great Lakes and the Ohio there is no more beautiful tract of country. At present, comparatively few of the citizens of Indiana are aware of its attractions; but it cannot long remain in obscurity. Among its hills and lakes thousands of the coming generation will find their summer homes.*

---

*A more detailed description of these lakes may be found in the 17th and 18th reports of the Indiana State Geologist.

VII.—THE NATURAL RESOURCES OF INDIANA.

W. S. BLATCHLEY, STATE GEOLOGIST

Too few of the residents of the state of Indiana have a proper conception of the natural resources found within her bounds. The text books on geography taught in the past, as well as those used at the present day, give but little exact information concerning those resources and that in a very condensed form. In the newspapers, which comprise the greatest educational factor of the masses, much has been published in recent years concerning natural gas, but while this resource has been of great value to a certain area of Indiana, it lacks much of being the most important natural resource of the state. Others there are, spread over a wider area, which have formed in the past, and will continue to form, sources of greater revenue and prosperity to the people at large. Of these, as well as of natural gas, some information will be given in this article, which, it is hoped, will prove of value to the teachers of the state.

The natural resources of the state of Indiana, as of any other restricted area of the earth's surface, may be classified into two great groups. The first of these consists of those forms of matter which have stored within themselves potential energy in the form of heat, which may be set free by combustion and then be controlled by some device of man and used by him to perform work. Such natural resources are called *fuels*, the most important of which, as found in Indiana are COAL, NATURAL GAS and PETROLEUM.

The second group of natural resources consists of those forms of matter which are devoid of any kind of stored energy which may be set free by combustion, but which are themselves used by man for varied and important purposes. The most valuable members of this group found in the state are SOILS, BUILDING STONES and CLAYS.

THE NATURAL FUELS OF THE STATE.

The fuels of the state, coal, natural gas, and petroleum, are valuable only for the stored energy in the form of heat which they contain. In speaking of these fuels the great law of the conservation and correlation of energy must ever be borne in mind. This law asserts, "That energy," like matter, "cannot be created, cannot be destroyed, but that one form can be changed into any other form." Man can invent no new forms of energy, nor can he produce a single iota of energy. He can only devise machines for transmuting or changing forms already existing into other and more available forms.

But the question naturally arises, how came this heat to be stored in the coal and other fuels? This question brings up another great truth which has become fully understood only in recent years; namely, that *the sun is the source of all the energy used in performing the work of the world.* From the sun comes energy in the form of heat and light which fall upon the grass and grain and trees of the earth, and furnish the power or force necessary for their growth. Heat and light enable plants to assimilate food and to grow, and at the same time energy is stored up in their tissues. Suppose, for example, that 1,000 calories (heat units) of heat are used in producing an ear of corn. When the ear is mature that amount of energy, no more, no less, is stored up in its substance. This energy can be made available to perform work for man in two ways: First, by burning the corn in a furnace, when the energy will be liberated again as heat and can be used to generate steam which in turn will cause wheels to revolve; second, by feeding the ear of corn to a horse, in whose body it will be changed into muscular energy which can be exerted in turning wheels or in pulling loads. Or, man himself can eat the corn, and the energy which is stored up within it will in his body be changed into muscular and mental energy. In other words, we move muscles and think thoughts with the energy derived from sunlight.

Plants alone have the power of thus storing up the energy of the sun's light and heat. Animals are wholly lacking in this power, and can utilize only the energy so stored by plants. This fact has been well portrayed by Professor Edward Orton in the following words:

"The remarkable office of the vegetable cell is thus brought to light. It is a storer of power, a reservoir of force. It mediates between the sun, the great fountain of energy, and the animal life of the world. The animal can use no power that has not been directly or indirectly stored in the vegetable cell. This storage is forever going on. Of the vast floods of energy that stream forth from the great center of our system, an insignificant fraction is caught by the earth as it revolves in its orbit. Of the little fraction that the earth arrests, an equally insignificant part is used directly in plant growth. But the entire productive force of the living world turns on this insignificant fraction of an insignificant fraction."

Bearing in mind this great truth, we can better understand how in ages past the sun's light and heat were locked up in the cells of those plants which flourished in the swamps of the carboniferous age. For thousands of years it accumulated within their stems and leaves and spores, and when, by the processes of nature, the plants were changed into coal it still remained, a most valuable heritage for future man. In the same way the energy stored up in the natural gas and petroleum of the Trenton rocks came from the sun and has been transmitted through the bodies of plants and animals.

The most important thing to remember in treating of these natural fuels is that they are not being formed in our state to-day. *No coal, no gas, no oil, is being made in Indiana by nature's processes, either in the bowels of the earth or above it.* Our present supply of each will never increase, but ever diminish. It is a great reservoir or deposit of reserve force upon which the people of the present generation are daily drawing without adding thereto. Like a bank account under the same conditions it is only a question of time until it will become exhausted.

COAL

Seven thousand square miles, or one-fifth of the area of the state of Indiana is underlain with coal. This area is found in the western and southwestern part of the state, and ranges from ten to sixty miles in width. It extends from Warren county southward 150 miles to the Ohio river, where it is widest in extent, stretching across the counties of Vanderburgh, Warrick, Spencer and part of Perry. Workable veins are found in nineteen counties in the area mentioned, and thin outcrops occur in three additional ones. At least seven distinct veins of workable thickness occur in the state. These vary from three to eleven feet in thickness, and aggregate in a few places from twenty-five to twenty-eight feet. The area of greatest development of the seams is embraced in the counties of Clay, Sullivan, Greene, Daviess and Pike; though Parke, Vermillion, Vigo, Owen, Warrick and Spencer rank as close seconds.

The coals of the state are of two varieties, which in places merge into one another. These are the non-caking or block coal and the caking or bituminous coal. The former is one of the most valuable fuels found in the United States. It has a laminated structure and in the direction of the bedding lines it splits readily into thin sheets, but breaks with difficulty in the opposite direction. It can be mined in blocks as large as it is convenient to handle, whence its common name of "block coal." It is remarkably free from sulphur or phosphorus, and when burning it does not swell out nor does it form a cake by running together. It leaves no clinkers, the only residue after combustion being a small quantity of white ashes. Ordinary bituminous coals have to have their volatile constituents driven off and be changed into coke before they can be utilized in the making of iron products. The sulphur which they contain, if allowed to remain, would destroy the tenacity and malleability of the iron. Their tendency to cake or become packed under the weight of the overlying mass in the blast furnace prevents the free passage of the heat through all portions of the molten iron. The block coal, on account of its freedom from sulphur and phosphorus and its non-caking properties, can be used without coking and thus becomes a most valuable fuel for the blast furnace and the cupola of the iron foundry.

For steam and household purposes it likewise has an unrivaled reputation. It burns under boilers with a uniform blaze that spreads evenly over the exposed surface, thus securing a more uniform expansion of the boiler plates. Its lack of sulphur also causes it to have but little detrimental effects upon the boilers, grates or fire-boxes. In household grates it burns with a bright, cheerful blaze like hickory wood, making a very hot fire, which for comfort and economy cannot be surpassed by any fuel except an abundant supply of natural gas. The block coal area lies mainly in Clay, Western Owen and Southeastern Parke counties, though small deposits are found in other sections.

The bituminous or non-caking coals found in Indiana vary much in purity and character, but their average will compare favorably with that of those found in any other state. They are far more abundant than the block coals, occupying an area of almost 6,500 square miles. Four workable seams are known, the maximum aggregate thickness of which is twenty feet, and the average aggregate thickness over the greater part of the district eleven feet.

The Indiana coal fields are as yet in the infancy of their development, yet last year, according to careful statistics gathered by the State Mine Inspector, Mr. Robert Fisher, 4,105,210 tons were mined from them. It has been computed that a ton of good coal used in a good engine will perform the same amount of work as 1,300 horses in a day of ten hours. The amount mined in Indiana last year had, therefore, stored up within it and capable of utilization, power or energy equal to that exerted by 14,621,300 horses working ten hours a day for an entire year.

The human mind cannot conceive the vast amount of energy at present locked up in the coal fields of the state, nor place anything like an accurate value upon it. The richest men of the nation to-day are those who have utilized the stored energy found in coal in years gone by; who have bought this energy at low prices, and either sold it in the form of manufactured articles at many-fold its cost price, or used it in transporting, for hire, man and his products to the four corners of the globe.

Natural Gas

During the past nine years natural gas has done more to advance the material interests of the state of Indiana than any other resource within her bounds. Millions of dollars of capital have been invested within the gas field, and thousands of people have flocked thereto, attracted by ready employment at good wages. As a consequence both the wealth and population of the area in which gas has been found have increased many-fold.

Originally that area embraced part of or all of seventeen counties lying northeast of the center of the state, and comprised on the whole about 5,000 square miles. On account of the encroachment of salt water and petroleum, this area has become gradually reduced until to-day the main gas field contains an approximate area of 2,500 square miles. This, however, is larger than has ever been possessed by any other state in the Union.

The average initial or rock pressure of the entire field in 1889 was 325 pounds to the square inch. To-day, according to careful measurements made during the past season by Mr. J. C. Leach, the State Natural Gas Supervisor, it is 230 pounds to the square inch over the main field. There is no doubt but that one-half of the entire supply has been nearly or quite exhausted, and as there can be no increase of it the pressure will decrease more rapidly in the future than in the past. How long the supply will last no man can tell. Too many varying factors, as the daily amount necessarily consumed for fuel and heat, the different pressures at which salt water and petroleum overcome the gas pressure—and more than aught else—the future percentage of waste, enter into the consideration of such a question. If the waste could be entirely shut off, the supply in the heart of the field, where much undeveloped territory has been held in reserve, would probably last for a number of years. It is in cities like Indianapolis and Richmond, which receive their supply through pipe lines, that the diminution in pressure is most noticeable; and there is no doubt but that their supply will become completely exhausted some time before that of the cities which lie wholly within the field.

PETROLEUM

Within the past two years the production of petroleum has attained enormous proportions in Indiana; the output for 1895 being 4,380,000 barrels.

The area in which the oil is found has steadily increased, and to-day comprises parts of Adams, Wells, Huntington, Grant, Blackford, Jay, Randolph, and Delaware counties. In addition, a few flowing wells are in operation in the city of Terre Haute, but repeated drilling has failed to locate any extensive field in Vigo county.

The probabilities are that the area of territory productive of oil will continue slowly to spread to the west and south until it finally embraces the greater part of the area at present yielding natural gas. This has been, in general, the history of other gas and oil fields, and there is no known reason why the one of Indiana should prove an exception. The oil, on account of its much greater specific gravity, underlies the gas in the area where the two are found together. As the pressure of the gas gradually decreases on account of a diminution of the supply, the hydrostatic pressure of the oil in time overcomes that of the gas, and a spouting or flowing well of oil results.

How much oil there is beneath the surface of Indiana is a question that no man can answer. How long it will last depends wholly upon the amount, and the average daily or yearly drain therefrom. Suffice it to say, *that the supply is limited, and will never be increased.* The age of a productive oil well in the United States does not generally exceed five years, and is often much less. A spouting oil well does not continue to gush forth for many weeks if allowed to flow freely. It soon degenerates into a flowing well and then into a pumping well, whose production dwindles away and finally ceases to be remunerative, so that unless new wells are continually being developed the output must fall off and finally cease entirely. However, there is no danger of the supply beginning to fail in Indiana for some years to come, as it has as yet been drawn upon for too short a time. True, some of the older wells have ceased to yield, but for every one so abandoned a dozen productive ones have been opened up; and this will continue to be the case until the total oil area, which can only be circumscribed by the future use of the drill, is fully developed.

Soil

Indiana is preeminently an agricultural state. Her soils constitute by far the most valuable of her natural resources. More people are dependent upon them for a livelihood than upon all the rest of her resources and manufacturing establishments combined. Ranking in area of square miles but thirty-fourth among the forty-five states of the Union, the census of 1890 shows that she stood fourth in the production of wheat, seventh in the production of corn, and eighth in the value of her live stock. This magnificent showing is due to two things: first, the excellent average fertility of her soils; second, the high degree of intelligence manifested by her farming population in the cultivation of the soils.

The soils of Indiana may be roughly classified into three great groups; viz., drift soils, residual soils and alluvial soils. The drift soils are found in the northern three-fourths of the state, are extremely varied in depth and character and are formed of a mass of heterogeneous material which was brought to its present resting place by a great glacier or slowly moving sheet of ice which, thousands of years ago, covered the area mentioned.†

The residual soils are found in the counties south of the southern limit of the glacier. They were formed, for the most part, in the place where they are now found, by the decay of the underlying limestone or

---

† The Glacial Boundary or southern limit of drift is shown on map p. 26.

sandstone rocks. The variety of materials entering into their composition is therefore limited, and they are, for that reason, among the poorer soils of the state.

The alluvial soils are those of the river and creek bottoms throughout the state. Gentle rains and earth-born torrents, little trickling rills and strong streams are ever at work tearing down the soils and underlying clays from every slope, and bearing them away to lower levels. The small water-formed trench of to-day next year becomes a chasm and ages hence a hollow, and the transported material is gradually deposited as alluvial soil over the so-called "bottom lands" which are annually overflowed.

In the production of any cereal nothing new is created, but forms of matter already existing in the earth, air and water are utilized by the growing plant. Taking wheat, for example, besides the carbon, hydrogen and oxygen, which make up the greater bulk of the straw and grain, and which are abundant enough in the air and water, potash, nitrogen, phosphoric acid, magnesia, lime, sulphur, chlorine and silicon are absolutely essential constituents. If any *one* of these is lacking in the soil, or is present in a form not available by the wheat roots, the plants will not flourish and the soil will be worthless for wheat production. Such a soil may, in most cases be made to produce a crop of grain by adding to it the constituent which is lacking, but if this can not be done except at a prohibitory cost, or one at which more fertile ground can be procured, the soil may be regarded as "worn out," barren.

The drift soils which cover the northern and central portion of Indiana, derived, as they were, from various primary and igneous rocks in the far north—ground fine and thoroughly mixed as they were by the onward moving force of a mighty glacier—are usually rich in all the above named necessary constituents of plant food. Neither they nor the alluvial soils require a large annual outlay for artificial fertilizers as do the residual soils of Southern Indiana over which the drift of the glacial period did not extend.

Building Stones

No state in the union possesses better stone for building purposes than Indiana. The oolitic limestone from Lawrence, Monroe and other counties has long been noted among architects for its strength and durability. It is of a uniform rich gray color and close texture, and on account of the ease with which it can be quarried, sawed and dressed for builders' use it can be sold with profit for a less sum per cubic foot than any other stone in America.

The best grades of it contain 98 per cent. of carbonate of lime, which is practically indestructible by ordinary atmospheric influences. It contains of iron oxide and alumina, two of the most damaging constituents of such stone, less than one per cent., thus showing a remarkable degree of purity.

The average crushing strength of twelve samples of tool-dressed oolitic stone, as determined by Major General Q. A. Gilmore for the Board of State House Commissioners in 1878, was 7,857 pounds per square inch, while that of four samples of sawed oolitic stone was 12,675 pounds per square inch.

The best deposits of the oolitic stone are found in a narrow strip of territory extending from Greencastle, Putnam county, to Salem, Washington county, a distance of 110 miles. The width of this strip varies from three to ten miles, and the stone throughout its full length is found very close to the surface.

Since the building of the court house and the state house at Indianapolis from this stone, its use for public and private buildings has steadily increased, especially in the East and South. A number of the private residences of the richer citizens of New York City have been recently constructed from it, while its use in such important buildings as the custom house at New Orleans, the Auditorium at Chicago, and many court houses in the counties of adjoining states has served to bring it more prominently before the attention of the public. During the year of 1895 the quarries in operation in the oolitic district had an output of more than 15,000,000 cubic feet, the most of which was shipped to points outside of Indiana.

Besides the oolitic limestones which are largely or almost wholly composed of calcium carbonate, numerous quarries are worked in the state in which magnesium carbonate is a leading constituent of the stone. These are found, for the most part, in the eastern half of the state, and belong to the Silurian formations. As a rule they are not building stone of a high grade. They are darker than the oolitic stone and are more apt to crumble after years of exposure, as they contain a larger percentage of iron oxide. The magnesia present also causes them to blacken and disintegrate more or less readily, especially in cities where the atmosphere contains large amounts of sulphurous fumes derived from the burning of soft bituminous coals. However, much of the magnesium limestone quarried in Indiana will compare favorably, both in texture and durability, with many of the stones used for building purposes in other states.

In a number of localities the sandstones comprising the so-called Conglomerate or Millstone Grit of Western Indiana, are valuable commercial stones, easily worked and of great durability. The Millstone Grit is a great formation lying at the base of the Coal Measure rocks of the state. In several places, notably at St. Anthony, Dubois county; Bloomfield, Greene county, and Portland Mills, Parke county, the stone is of a handsome brown color, and compares favorably in appearance with the brown sandstones of the Lake Superior region which are so much used for the fronts of business blocks in Chicago, Milwaukee and other cities of the northwest.

One use to which this brown sandstone is peculiarly well fitted is for the lintels and cornices above the windows and doors of those buildings whose fronts are composed of dry pressed brick. Where limestone is used for the lintels, the rain, dashing against it, is sure to dissolve out a small portion of the stone which flows down over the brick and gives them a mouldy, streaked appearance. Where the brown sandstone is used no such streaking is seen on the brick beneath the windows and archways. The color of the sandstone also harmonizes better with that of the brick than does that of the limestone.

Large quarries of buff and gray sandstone have also been in successful operation at Attica, Williamsport, and Riverside, and at Cannelton, on the Ohio river.

CLAYS

Among the most valuable of the undeveloped resources of the state are her clay deposits. In one form or another they are found in every county, but the largest and most valuable ones occur in the western and southwestern parts of the state, where the coal measures exist, for the coal measures of the state are preeminently its clay measures. Every seam of coal is normally underlain with a bed of fire clay, and above the coal there are almost always beds of shale.

These coal shales a few years ago were thought to be worthless, but experiment has proved that they are excellently adapted to the making of paving brick, roofing tile, sewer pipe, and many similiar products. In Ohio, where forty-four paving brick factories turned out 208,000,000 paving brick in 1894, 80 per cent. of the best grades were made of the carboniferous shales which ten years ago were wholly unused.

The fire-clays from beneath the coal can be utilized for the same purposes as the overlying shales, and many of the better grades can be made into refractory wares of good quality. Some of them are also well suited for potter's use.

In Lawrence, Martin and Owen counties there occur large deposits of a white kaolin, which is the highest grade clay found in the state. A careful chemical analysis shows that it contains less than two per cent. of impurities. The quantity of iron oxide is so small as to have no effect upon the color of the wares made from it, they being, if anything, whiter than the clay itself.

Like many similar kaolins, this is practically non-plastic; but by grinding very fine and then kneading, it can be made to assume a certain degree of plasticity. Its refractory properties are of the highest, and mixed with a small percentage of a more plastic material, as one of the purer underclays of the coal seams, it can be used in the making of the finer grades of retorts, glass-pots, glass-tanks, etc. Ground fine and pressed dry it will make the highest grade of fire-brick. Much of it has been utilized in the past for the making of pottery, and also for alum salts used as sizing for the finer grades of writing and wall paper. Thousands of tons of this purest of clays are visible in the mines which have been opened near Huron, Lawrence county, but at the present time it is not being worked. The stratum thickens as progress is made further back into the hills. The deposit is not a local one covering a few rods or acres, but square miles, as evinced by outcrops which are known. There is enough in sight in the mines at this one deposit to last an average factory a hundred years, and not one-thousandth of it has been exposed to view. There it lies, a great mineral resource of untold value, unworked, unutilized, awaiting only the coming of energy and capital to make it up into many kinds of products which are now brought into our state from distant lands.

The clay-working industries of the state have grown apace in the last five years. No hesitation is felt in prophesying that within the next ten years they will become the leading manufacturing industries of Western Indiana. Raw material and fuel, both of excellent quality, are found associated together in enormous quantities in many places which are accessible to transportation; and where the three elements of fuel, raw material and railways are thus combined, capital in time is sure to locate and utilize the natural resources. The larger clay industries already in existence in the vicinity of Brazil, Terre Haute and other places, are all of them flourishing; the demand for their products in many instances being greater than the possible supply. They have proved by practical experience that the shales and under-clays of the coal measures are in every way fitted for manufacturing purposes.

One of the chief beneficial effects which the development of the clay-working industries will bring about, will be the increasing of the available amount of coal in the state. Many seams now thought to be too thin to work will be utilized in connection with the associated shales and fire-clays. The minimum thickness of a workable seam of coal will, therefore, be greatly reduced, and many veins which have long been allowed to pass unnoticed will be mined with profit.

Iron Ores

Limonite or bog iron ore, and siderite or kidney iron ore, are found abundantly in several counties of Indiana. The former occurs, notably, in Greene, Martin and Perry counties, and in the swamps of the Kankakee region—the siderite in all of the coal-bearing counties. Experience has proved, however, that these ores are too silicious to compete with the rich beds of hematite of Missouri, Tennessee and Georgia. As a proof of this, it is only necessary to state that of fourteen blast furnaces which have been erected in the state in the past, not one is in operation at the present. Most of them have long since gone to ruin, and of those still standing the last one went out of blast in 1893.

OTHER MINERALS AND ORES

With the exception of small quantities of drift gold in the form of minute grains and scales, which are found in the sands and gravel beds along the streams of Brown, Morgan and other counties near the southern limit of the drift area, no gold, silver or other precious metals occur in the state. Much money has been foolishly spent, and time wasted by people who have thought otherwise; but they have ever had their labor for their pains.

In many of the northern counties small pieces of "black-jack" or zinc blende, galena or lead sulphide, and native copper, are occasionally found, and give rise to much local excitement and speculation. It is needless to say that the specimens of lead and copper were brought in by the ice-sheet or by the Indians; and the blende, while possibly of native origin, is utterly valueless. In almost every county one also hears tales of reputed silver and lead mines, which in the days of long ago were secretly worked by the Indians. Many well-informed people yet believe these tales, and have spent days in fruitlessly searching after imaginary mines, where enough silver may be had to pave the streets of their native towns, or where lead ore exists without limit.

While Indiana is thus lacking in the precious and other useful metals, her deposits of coal, clay, stone, natural gas and petroleum are far more valuable, and will in time bring more wealth into the state than if, instead of them, rich mines of gold and silver had been found within her bounds. Higher grades of labor and more stable industries are based upon such resources, for few, if any, large factories utilize gold and silver in quantity as a manufacturing resource.

VIII.—INDIANA : A CENTURY OF CHANGES IN THE ASPECTS OF NATURE *

AMOS W. BUTLER

It is probable that the first white man within the boundaries of Indiana was the explorer, LaSalle. His voyage was made about 1669. The earliest settlements were established within the first quarter of the last century at Ouiatenon and Vincennes. Authorities do not agree as to which was settled first or the date of the settlement. These were but trading posts. Their effect upon existing conditions was but small. Nor was it until the English began to occupy this region, at the opening of this century, that the old began to fade before the new.

FORESTS

Over the greater part of this state were spread dense forests of tall trees—heavy timber—whose limbs met, and branches were so interwoven that but occasionally could the sunlight find entrance. There was little or no undergrowth in the heaviest woods and the gloom of these dense shades and its accompanying silence was terribly oppressive. Mile upon mile, days' journey upon days' journey, stretched those gloomy shades amid giant columns and green arches reared by nature through centuries of time. The only interruption were the beds of water-courses; the poorer hill-sides covered with underbrush; the smaller growth of less productive uplands; the site of an extensive windfall, the record of a tornado's passage; the small area of second-growth timber marking the former clearing for some Indian camp; the more or less extensive patches of meadow, the result of the destruction of the forest by Indian fires. To the west in the valley of the Wabash, were wide meadows covered with long grass. In the northern third of our territory were prairies and sloughs alternating with wooded sand-hills and reedy swamps, imperfectly drained by a network of sluggish streams, which, in turn, gave place to extensive marshes toward Lake Michigan.

The southern portion of the state was more heavily timbered. Perhaps nowhere could America show more magnificent forests of deciduous trees, or more noble specimens of characteristic forms than existed in the

* Address of the President of the Indiana Academy of Science, at Indianapolis, Dec. 27, 1895.

valleys of the Wabash and Whitewater. The trees decreased in size to the northward, those along the lakes being noticeably inferior. Coniferous trees were few in number and confined to restricted areas. Those found were poor representatives of their species. The forests were made up of many kinds of trees growing together indiscriminately. Here and there a certain group or occasionally a species was found predominating. In various localities the character of the forest was different. While oak, ash, hickory, maple, beech and elm were prevailing trees, they varied much in number and proportion. In some places the tulip poplars were very numerous, the trees often attaining great size—the largest tree of the primitive forest. Forty-two kinds of trees in the Wabash valley attained a height above a hundred feet; the tallest recorded being a tulip poplar 190 feet high. It was twenty-five feet in circumference and ninety-one feet to the first limb. Many thousands grew over the state measuring from three and a-half to ten feet in diameter. Numbers of sweet gum in the more fertile ground in the southern part of the state contended with the tulip poplar in height, and in beauty and symmetry exceeded it. They sometimes attained a height of 150 feet and a diameter of four feet, often preserving almost the same size to the first limb.

In the oak woods there were giants, too. The red, scarlet, burr and white oaks reaching a girth of ten to twenty feet, and often a height of 125 to 150 feet. One instance is reported of a scarlet oak 181 feet high. In the southern part of the state, too, the sweet buckeye attains great size, often being three and a-half to four feet in diameter, with a trunk as straight as a column, and reaching a total height of over 100 feet. One example of this species is unique. It is the tree from which was made the celebrated buckeye canoe of the Harrison presidential campaign of 1840. The tree grew in the southeast corner of Rush county, and is said to have been, when standing, twenty-seven feet nine inches in circumference and ninety feet from the ground to the first limb. Here and there, quite thickly scattered, were to be found groves of the finest black walnut trees the world has ever known. Some of these groves were quite extensive, containing hundreds of trees, individuals of which were four to six feet in diameter and 100 to 150 feet high.

In the river valleys, along the streams, the great size of the sycamore was noticeable. This was the largest of the hardwood trees, reaching a maximum height of 140 to 165 feet and often measuring five to ten feet in diameter. Keeping these company were the cottonwoods, the larger of which measured five to eight feet in diameter and 130 to 165 feet high. The beauty of all the trees of this region was the white elm. Its diameter was three to five feet and its height sometimes 120 feet or more, the ambitus often spreading over 100 feet.

Indian Villages

At the time of its settlement the southeastern third of our territory, including all the Whitewater valley, contained no towns, and was unoccupied by the Indians save as occasionally a hunting or a war party passed through it. In the valley of the Wabash and in the northeastern part of the state were Indian villages, located because of natural advantages. These have been apparent to the whites, who in several instances established upon their sites, settlements, some of which have since become prominent as towns or cities. Kekionga (Fort Wayne), Chip-kaw-kay (Vincennes), and Ouiatenon, on the west side of the Wabash river, four miles below Lafayette, were selected as trading posts by the whites, being centers of the finest game region occupied by man within the limits of the present state. The peltry from the last mentioned post in one year, in those early times, amounted to about 8,000 pounds sterling.

Animals

In different localities under different conditions were different forms of life. We have noted this concerning plants. It was so regarding animals. American bisons, buffaloes they are generally called, ranged in countless numbers over the meadows and prairies at the time we first learn of them. The Whitewater and Miami valleys formed routes to the Ohio river and the Big Bone Lick in Kentucky. The Wabash valley became another avenue for their journeys, and the old trail from the prairies of Illinois to the Kentucky barrens, crossed the Wabash river below Vincennes. Over this wide, well-marked road, evidences of which still remain, countless thousands of bisons passed annually. From the Ohio river to Big Bone Lick was a road spacious enough for two wagons to go abreast, made by these animals. Evidence of their former abundance is preserved in the swamps about this lick. In places their bones are massed to the depth of two feet or more as close as the stones of a pavement, and so beaten down by succeeding herds as to make it difficult to lift them from their beds. The Blue Licks in Kentucky was also a favorite spot. In 1784 Filson says: "The amazing herds of buffaloes which resort thither, by their size and number, fill the traveller with amazement and terror, especially when he beholds the prodigious roads they have made from all quarters, as if leading to some populous city; the vast space of land around these springs desolated as if by a ravaging enemy, and hills reduced to plains: for the land near these springs is chiefly hilly." In the region that was densely wooded the bisons were only seen as transients, but in the meadows and prairies they abounded. From the summit of the hills at Ouiatenon we are told that in 1718, nothing was visible to the eye but prairies full of buffaloes.

Elks were common and deer still more so. Bears and wolves were

A CENTURY OF CHANGES IN THE ASPECTS OF NATURE 75

quite abundant. In one favorite locality, it is reported, a good hunter, without much fatigue to himself, could supply daily one hundred men with meat. Beavers were found in many localities. Especially favorable to them were the more level regions to the northward. Otters were quite common, while wild cats, Canada porcupines and panthers were numerous.

Of snakes especially noticeable for their abundance were rattlesnakes and copperheads.

The ponds, sloughs and deeper swamps were the homes of many species of fishes, mollusks and crustaceans. The creeks, shaded by the closely-crowding trees, contained water all the year round, and in them smaller fishes reared their young. The rivers were clogged and dammed with fallen trees and driftwood, and the water, when the streams were swollen by heavy rains, pouring over these obstructions, cut deep holes which became the homes of great numbers of larger fishes.

Wild turkeys were found in large flocks. Bob-whites were so numerous that when they collected in the fall as many as a hundred were taken in a day with a single net. Ruffed grouse were abundant. Ducks and geese, snipe and plover were found in inestimable numbers where favorable conditions existed. Paroquets were more or less numerous over the entire region, and in the lower Wabash and Whitewater valleys were as abundant as blackbirds are in spring and fall. Passenger pigeons bred and roosted in many localities. During the migrations they appeared in such numbers that they obscured the sun and hid the sky for hours, sometimes for days in succession. The strange appearance was made more wonderful by the continuous rumble of the thunders of the oncoming clouds—the noise of the strokes of millions upon millions of wings.

Besides these, more rarely swallow-tailed kites and ivory-billed woodpeckers added their characteristic forms to the wild scenery along the Ohio river. The osprey and bald eagle built their nest beside the streams, and while one fished the other plundered the fisher.

Within the dense shades of the deeper woodland there was but a small number of birds. There, quiet reigned; twilight by day and densest darkness by night. How oppressive the awful quiet amid those gloomy solitudes! Everywhere the smaller birds were few compared with their present numbers.

Results of White Occupation

But men of our race came upon the scene. Indians there had been before. As it always has been and so will continue to be when two races, one superior, the other inferior, come into competition the superior will overcome. The contest was unequal. The barbarism of the Ohio valley could not hold its own against the alert and thoroughly equipped pioneer. Soon the native began to part with his land. It was not long until many sought other homes. Others attempted to become permanent residents and to adopt, in some measure, the habits of the conquerors. The result is too well known. An ancestor of theirs, gifted with the power of a seer, may have been the subject of these lines:—

> "There was once a neolithic man, an enterprising wight,
> Who kept his simple instruments unusually bright;
> Unusually clean he was, unusually brave,
> And he sketched delightful mammoths on the borders of his cave.
>
> To his neolithic neighbors, who were startled and surprised,
> Said he, 'My friends, in course of time we shall be civilized!
> We are going to live in cities and build churches and make laws;
> We are going to eat three times a day without the natural cause;
> We are going to turn life upside down about a thing called gold;
> We're going to want the earth and take as much as we can hold.
> We're going to wear a pile of stuff outside our proper skins;
> We're going to have diseases! and accomplishments! and sins!!!'"

In the office of the Secretary of State, at Washington, is an impression of "The seal of the Territory of the U. S., N. W. of the Ohio River." Of this the late William H. English in his "Conquest of the Country Northwest of the Ohio River," says: "A study of this historic seal will show that it is far from being destitute of appropriate and expressive meaning. The coiled snake in the foreground and the boats in the middle distance; the rising sun; the forest tree felled by the axe and cut into logs, succeeded by, apparently, an apple tree laden with fruit; the Latin inscription '*meliorem lapsa locavit*,' all combine to forcibly express the idea that a wild and savage condition is to be superseded by a higher and better civilization. The wilderness and its dangerous denizens of reptiles, Indians and wild beasts, are to disappear before the axe and rifle of the ever-advancing western pioneer, with his fruits, his harvests, his boats, his commerce, and his restless and aggressive civilization.

Meliorem lapsa locavit!

He has planted a better than the fallen!

The white man made the navigable waterways his routes and settled along them. At once, under his influence, the aspects of nature began to change. As in every other land the effects of man's settlement began to be seen. The need for food and clothing and the desire for tillable land were the great causes which impelled him to action. In every land, on every sea, the story has been the same. Before his aggression disappeared the most noticeable forms of life. The large or conspicuous species are those most easily affected, the ones which first are destroyed. The disappearance of the great animals of Europe, of the bison and the urus; the extinction of the giant birds of New Zealand, of Steller's sea cow and the great auk, one each upon our eastern and western coast; the most wonderful destruction of the great herds of American bison, the threatened extinction of the fur seal in the North Pacific; of the zebra, camelopard and other large animals in Africa, are but notable illustrations of the greater changes that have been wrought. There are smaller ones not so conspicuous but more potent in their influences upon human welfare.

The bison, the most characteristic of all the animals of America, was the first to disappear from the region under consideration. Formerly it had ranged east, at least as far as Western New York and Pennsylvania and in Virginia almost to tide water, but in 1808 it was exterminated east of the Wabash. The elk followed it closely, disappearing from the Whitewater valley about 1810, and from the state in 1830. The panther followed soon after. Virginia deer, bear, wolf, otter, beaver and other forms were almost exterminated, though of some, if not all of these latter forms, a remnant yet remains in some favored localities.

Turkeys and bob-whites; ivory-billed woodpeckers and wood ibises; black vultures and Carolina paroquets have been almost, or in a great measure, exterminated. The paroquet which ranged to the Great Lakes and were so common a feature in the landscapes of pioneer times have not only disappeared from Indiana, but from almost all the great range from Texas to New York, over which they spread at the beginning of this century, and are, perhaps, now only found in a restricted area in Florida. The day of their extirpation is near at hand.

The passenger pigeon survived the beautiful little parrot until a later day. But nets and guns, a short-sighted people and inefficient laws have all but swept out of existence this graceful bird. It is now on the verge of extinction. We can no more appreciate the accounts given of the innumerable hosts of these birds of passage than we can of the incalculable multitudes of the bisons three-score years ago. The words of those who saw them, we are assured, do not in any way convey an adequate idea of the wonderful sights and sounds during a flight of pigeons. Some of their roosts covered many miles of forest. There, as they settled at evening, the gunners from near and far began to collect for the slaughter. The loaded trees upon the borders of the wood were first fired upon. Then they passed into the denser forest. Three or four guns fired among the branches of a tree would bring down as many as two bushel sacks of dead birds, while numbers of cripples fluttered beyond reach. After a number of shots over a considerable area, several acres perhaps, the whole roost would rise with a deafening thunder which no one has attempted to describe, and soar out of sight in the dusk of the early evening while from the rising cloud came a noise as of a mighty tornado. As the darkness settled the birds descended and alighted, many deep, upon the branches of the trees, the weight being sufficient to break off many of the large limbs. Then the scene changed. The slaughter began in earnest. The rapid firing of guns; the squawking of the pigeons; the breaking of the limbs of giant trees beneath their living weight; the continuous rumble arising from the whirr of countless wings; all illumined by the lurid lights from numerous torches and many fires, produced an effect of which no words can convey a conception to one who has not experienced a night at a pigeon roost. Each year such scenes were re-enacted. Each year the slaughter went on. Less and less the numbers grew. Trapping and netting, supplemented by repeating guns, added to the power of destruction, and the pigeons whose numbers were once so great that no one could conceive the thought of their extinction, have dwindled until they are rarely found. One pigeon in a year! Soon they will be but a memory.

Removal of Forests

The pioneer's first work was to cut away the trees and build a cabin. As each cabin was built it foreshadowed a clearing extending more and more each year. The line of the Ohio and the Wabash formed the basis for the advance of settlement. The axe and the fire performed their work. Great deadenings gave promise of a lively time log-rolling next season. Giant tulip poplars, monster black walnuts, and oaks, ash, wild cherry and sweet gum, the largest of their fellows, were rolled into heaps and burned. To this, in time, was added the demand for fuel, for lumber and for timber to supply all the drafts which human wants could make upon the forest, not only for our own population but for other states and other lands. Thus were our forests destroyed. Now, except in a few localities, there remains no virgin forest.

The destruction of the primitive woods costs much besides the trees that were sacrificed. Each tree was the nest or resting place of other forms of life—of the blight upon its leaves; of the fungus upon its limbs; of the lichen and moss upon its bark; of the birds among its branches; the insects on its foliage and about its blossoms; the borers within its body. And it sheltered other lowly ground-inhabiting forms beneath its spreading shades. Who can tell what the destruction of a tree signifies? How far-reaching are its effects? After the axe came fire, carrying destruction to the more inconspicuous animals and plants. Fire, too, swept the standing woods and its blighting effects extended far beyond the immediate necessities of the pioneer. With the cutting away of the larger trees in many localities sprang up thickets, and therewith came thicket-inhabiting animals. As the clearings were extended meadow lands and pasture lands were reserved. To the meadows came such forms as the bay-winged bunting, field sparrow, grasshopper sparrow, meadow lark, meadow mice, garter snakes, green snakes, bumble-bees and grasshoppers—species peculiar to such surroundings. Some parts of this land were wet, and where the drainage was poorest became swamps and sloughs. There, forms which love such places came. Among them marsh wrens, swamp sparrows and red-winged blackbirds, salamanders, frogs, water snakes, aquatic insects and marsh plants. The orchard and garden developed, birds well known to us and greatly beloved for their cheery social ways, there made their home and lived upon food brought to the locality by the changing conditions. The number of settlers increased, causing steady diminution in numbers of all the larger mammals, especially those used for food or valuable for fur; of geese, ducks and other water-loving birds. The early settlers had brought with them the black rat. Later another form, the brown rat, which like the first was native of the old world, appeared, following the routes of civilization. It drove out the other rat and has since occupied its place. The shy gray fox disappeared in advance of the oncoming pioneer, and the red fox occupied the field left vacant. The hog, a most valuable factor in the development of the West, proved equally valuable as an ally in the warfare against snakes. Largely through its efforts were the rattlesnakes and copperheads destroyed.

Removing the timber and breaking the ground began to show its effect upon springs and water-courses. Many became dry during the warm season. All life, be it salamanders, fishes, mollusks, insects or plants that found therein a home, died. As time went on drainage became a feature introduced into the new country. With the draining of the sloughs and swamps other changes came. The birds that lived among their reeds and flags, mingling their voices with those of the frogs, disappeared, and the land reclaimed tells, in its luxuriant growth of corn, no story to the casual passerby of the former population which occupied it.

And so it was. Change succeeded change; little by little, but still each cleared field, each drained swamp, each rotation of crops, each one of a thousand variations in cause had its effect upon the numbers and life-histories of our plants and animals.

When the Indians left, the prairies were no longer annually burned over. Forest vegetation began to seize upon this open land and, in time, much of it became reforested. Into it was brought life from the surrounding woods and the former occupants were driven out.

With the thinning of the trees in the woodland appeared an undergrowth. Where the undergrowth came, and where the second-growth appeared in neglected clearings, the vegetation was often different from that of the original forest. This, too, was destined to go the way of passing things. The ginseng, spikenard, bloodroot, yellowroot, mayapple and many ferns are following the woody plants to extermination. Milk sickness, once so prevalent among the early settlers, with the peculiar fevers of the new country, are of the past. Staggers has disappeared from many places, yet the wild larkspur, which traditionally is its cause, has become more abundant in some congenial localities, and in such neighborhoods the disease is quite serious.

But there are other results of the introduction of civilization which have made themselves felt. The streams were dammed and the migratory fishes prevented from ascending them. The driftwood disappeared from the water courses. In time the dams, too, were gone. The deep holes, where the fishes loved to hide, filled up. The streams carried less water through the summer. Dynamiting, netting, and other illegal means of fishing became prevalent. All these have combined to wage a war of extermination against the inhabitants of our streams and lakes which might, if properly protected, prove an exceedingly valuable factor alike in the enjoyment and in the food supply of our people.

The telegraph wire is very destructive to birds. Birds and insects have found a new instrument of destruction in the electric light. Many living things besides man have found that railroad tracks are dangerous. They, in turn, are highways along which the cars introduce new forms of plant and animal life. The self-binder and the mower play havoc with the lives of many inhabitants of the medows and grain fields.

Following in the civilizer's footsteps have come other changes. Man has not only made the wilderness to blossom as the rose, and gathered fruits and grain from all lands for the necessity and enjoyment of our people, but with the grain has been sown tares, and with the fruit has been planted blight. Teasels, Canada thistles, wire grass, plantains and prickly lettuce are contending for the soil. Pear blight, black knot, smuts and rusts affect fruits and flowers. Chinch bugs, Hessian flies, Colorado potato beetles, clover-root borers, scale insects and cabbage worms dispute with the farmer his right to the crops he has planted.

Some of the native forms of life have changed their habits in some respects. This is evidenced by the rose-breasted grosbeak feeding upon the Colorado potato beetle, the destruction in the rice fields of South Carolina caused by the rice birds—our bobolinks, the loss inflicted in the rice swamps of Louisiana by the red-winged blackbirds, the damage done to the western corn grower by the bronzed grackle—our common blackbird.

By man's agency the European house sparrow, or English sparrow, was introduced and as its numbers increased it began to assert itself in the struggle for existence. The bluebird, which had come from her hole in the snag, was driven from her box. The martin, which, like the chimney swift, formerly nested in hollow trees, left its nesting sites about the house. Even the cave swallow, which in olden times fastened its nests to the cliffs, was, in some cases, driven away. The warfare with this aggressive little foreigner still continues, worse in some places than others. But it has such surprising powers of reproduction and unheard of audacity it seems that it must soon cover our entire continent. The history of the German carp in this country illustrates the same persistent and successful struggle for the mastery in our waterways that has been noted of the house sparrow on the land.

In time, fashion demanded that which neither man's appetite nor his need for protection had impelled him to take. Her altars were erected and upon them sacrifices—a host innumerable—were offered. Fur-bearing animals and bright-plumaged birds were most earnestly desired, but even the shells of turtles, the skins of snakes, the teeth of alligators and the pearls of fresh-water mussels were acceptable offerings. The extent of the destruction of innocent bird-lives alone is appalling. A few facts may convey some idea of this. Among the items of one auction sale in London were 6,000 birds of paradise; 5,000 Impeyan pheasants; 360,000 assorted skins from India; 400,000 humming birds. One dealer in 1887 sold no less than 2,000,000 bird skins. From information obtainable it is certain that hundreds of thousands of birds must have been slain in the United States for the glory of fashion's devotees. It is probable that not less than 5,000,000 birds were required each year to supply the demand in this country alone when the bird-wearing craze was at its height. To this great number of victims our own state has been to a greater or less extent, a contributor. Many counties in Indiana were visited by bird hunters. It is said from Indianapolis alone 5,000 birds prepared for millinery purposes were shipped in one year. Under our present law, which seems to be well enforced, it is a pleasure to say, our birds are apparently free from that danger.

Changes still continue. The future will record them as has the past. Those to come promise to be more fruitful of results; to be of greater moment to mankind; to bring more earnest messages for human weal or woe. But during no time in the future will the changes in the aspects of nature here be so notable, so incomprehensible because of their vastness, as have those of the century just closing.

IX.—A STUDY OF THE CITY OF TERRE HAUTE

CHARLES R. DRYER

While cities are the most artificial features of the earth, yet they are most intimately dependent upon natural conditions. They are the hieroglyphics which record in summary and epitome the whole geography of a country, physical, political and commercial. The student should be taught to read that inscription. The main lines of investigation are two:

1. What conditions originally determined the location of the city?
2. What causes have conspired to make it a large city?

These problems should be stated in terms intelligible to the students and they should be set to work to find solutions for themselves. *Don't tell them anything.* Some boy may make the brilliant discovery that rivers have a strong inclination to flow through large towns; and that discovery is worth more *to him* than all the information about it that could be crammed into him. The young cities of America afford much easier work than the cities of the old world. The investigation may be conducted in regard to one particular city, as Chicago, or to the cities of a certain area, as New York, England, the Mediterranean shore.

Nearly all important cities will be found to have an intimate relation with water routes. The sea and the river are the natural highways of the world. The cities which are situated on bays or inlets which afford good harbors and easy communication inland, on deltas, at the head of tide water, at the head of navigation, at rapids, falls, or divides which necessitate a "carry" or portage of goods, or at an easy stream-crossing, ford, ferry or bridge, include a very large per cent. of all; and the student should be led to discriminate between these various conditions. If to these be added the cross-roads cities, which stand at the intersection of great routes of land travel, there will remain but a few special cases to be accounted for otherwise. Let cities be classified according to these characters; but *let the student do his own classification*, which means only that he should be permitted, encouraged, compelled to do his own thinking.

In city schools the same plan of inductive study may be applied to the home city with great profit. Let the students investigate the origin and growth of their own town, how a certain street or quarter came to be devoted to business, another to manufacture, and another to residence. Let them learn how their own city is managed and governed, about paving, street cleaning, sewerage, gas and water supply. This is geography of the most intensive educational, scientific and practical value.

OUTLINE FOR A STUDY OF THE CITY OF TERRE HAUTE *

I. CONDITIONS WHICH DETERMINED THE LOCATION OF THE TOWN

1. Preliminary Events.

Occupation by Mound Builders——occupation by Indian Tribes——French discovery and occupation——posts at Vincennes and Ouiatenon (Lafayette)——cession to the English——Clark's Conquest——cession to the United States——ordinance of 1787——organization of the Territory of Indiana——seat of government and land office at Vincennes——Indian wars and extinguishment of Indian title to land——Fort Harrison——government survey——decline in importance of Vincennes.

2. Physical Conditions.

Wabash river——navigability——good landing place——straight reach of river——unbroken river front——flood plain on east side narrow——Macksville terrace narrowed bottoms and made easy crossing——broad valley above and below——extensive prairies and bottom lands——gravel terrace——wooded ridge parallel with and near river——good drainage.

3. Founding of the Town.

French settlement at "Old Terre Haute" three miles below and English settlement at Fort Harrison were rivals——a town mid-way between would kill both——site of old Indian village——junction of Louisville, Vincennes and LaFayette roads——land owned by the founders——who were the founders?——motives in founding——origin of name——original plat——early settlers——sources of subsequent immigration.

II. INFLUENCES AND CONDITIONS WHICH LED TO THE GROWTH OF THE TOWN

Navigation on Wabash River——organization of Vigo County and location of county seat——the National road——the Wabash and Erie canal——canal to Evansville——corn and wheat lands——milling——pork packing——distilling——blue grass pastures——horse raising and

---

* This outline was prepared at the request of Principal Charles S. Meek, of the Terre Haute High school, for the use of his senior class. For assistance in its preparation, acknowledgment of indebtedness is made to an article on the Evolution of Cities, by the eminent French geographer, Elisée Reclus, in the Contemporary Review for February, 1895, an abstract of which appeared in THE INLAND EDUCATOR, Vol. I, p. 310; to the paper of Superintendent W. B. Powell, of Washington, D. C., read at the International Geographic Conference at Chicago, and published in the National Geographic Magazine, Vol. V, p. 137; to a plan for a study of the Rural Community, by Superintendent W. A. Millis, of Attica, Ind., THE INLAND EDUCATOR, Vol I, p. 298; and to Dr. J. T. Scovell of Terre Haute, for free use of his intimate knowledge of the locality.

racing——native timber, walnut etc.——the railroads, T. H. & I., C. C. C. & St. L., E. & T. H., etc.——coal mining——blast furnace and rolling mill——brick and tile manufacture, etc.

III. THE MAKERS OF TERRE HAUTE

Bullitt, Markle, Lasalle, Aspinwall, Early, Griswold, Reed, Dewees. Gilbert, Cruft, Linton, Farrington, Ross, Warren, Modesitt, Rose, McKeen, Deming, Kidder, Hulman, Collett, Crawford, etc.

The study of contemporaries may be omitted or continued at discretion.

IV. STAGES OF DEVELOPMENT

Growth in space——causes of extension in various directions——growth in population——growth in value of property——changes in character of buildings——changes in character of population——changes in social conditions——changes in sanitary condition——disappearance of malaria——growth in schools and churches——changes in municipal government——incorporation of town or city.

V. THE PRESENT CITY

1. Advantages and Disadvantages of Location.

Latitude and longitude——climate——mean temperature for January and July——absolute range of temperature——amount and distribution of rain-fall——distances from sea board, Great Lakes, Ohio river, Mississippi river, Gulf of Mexico——distances and relation to other towns and cities——relation to surrounding country.

2. Advantages and Disadvantages of Site.

Topography——subsoil drainage——wells and cess-pools——facilities and difficulties of sewerage——access easy from three directions, difficult from the west——abundant supply of gravel for streets——abundant supply of clay for brick——absence of water power.

3. Plan of the City.

Boundaries and area——direction and location of streets——area of blocks and lots——width of streets and sidewalks——name system of streets——number system of houses.

4. Distribution.

Business districts——causes of location——tendency away from the river——manufacturing districts, causes of location——best residence districts, causes of location——good residence districts, causes of location——poor residence districts, causes of location——slums, causes of location——suburban districts, causes of location——value of property in each district——relation of manufacturing districts to health and comfort of residence districts.

5. *Buildings.*
Materials——size——regulations concerning——cost of materials——methods of construction——architecture.

6. *The People.*
Number——density——race——sex——age——occupations——married and single——death rate, etc.

VI. THE MUNICIPALITY

Common Council——Mayor——Clerk——Treasurer——how elected——powers——duties——the city charter, compare with those of Indianapolis, Ft. Wayne, etc.

VII. TRAVEL AND TRANSPORTATION

1. *Material and Condition of Streets.*
Pavements——asphalt, brick and other——cost and relative value——sidewalks——flag, cement, brick and other——cost and relative value——rules and ordinances for construction and maintenance——bridges——amount of travel across——street cleaning——how done——why is snow seldom removed?

2. *Vehicles and Passengers.*
Character of vehicles——drays, trucks, omnibuses, cabs, etc.——number passing various points——number of foot passengers passing various points——rights and privileges of foot passengers and vehicles.

3. *Street Railways.*
Franchises——how obtained——terms——value——routes——length of track——number of cars——speed——frequency——capacity——number of passengers carried——motive power and mechanical system——cost of plant——maintenance and running expenses——cost per car mile——profits——class of people using——length of average ride——time of day of greatest travel——fares——difference between five and three-cent fares and its relation to income and wages——rights of public to use of street——rights of railway companies to use of streets——municipal control or ownership of street railways——compare Detroit, Toronto, Glasgow.

4. *Relation of Trunk Line Railroads to City.*
Stations——freight house and yards——crossings——number of passenger trains——number of passengers——number of freight trains——amount of freight shipped from and delivered to the city——express companies and express business.

5. *Main Wagon Roads Leading into the City and Amount of Travel on Each.*

6. *Transportation on Wabash River.*

VIII. COMMUNICATION

1. *Telegraph and Telephone Companies.*
Service——rates——use of streets and alleys by.

2. *Mail Facilities and Delivery.*
    Amount of mail delivered and sent——income of office——organization.
3. *Municipal or Governmental Control of Telegraph and Telephone Service.*

IX. POLICE DEPARTMENT

Officers——Courts——appointments——responsibility——crimes——arrests——convictions——sentences——enforcement of law.

X. FIRE DEPARTMENT

Officers——employes——companies——engine houses——engines——alarm boxes——equipment——efficiency——cost——number of fires——ability to deal with a great conflagration.

XI. ENGINEERING DEPARTMENT

Employes——appointments——duties——importance of.

XII. PUBLIC HEALTH

1. *Board of Health.*
    Appointment——powers——regulations——infectious diseases.
2. *Disposal of Sewage and Garbage.*
    Extent to which cess-pools and vaults are used——dangers——extent to which sewers are used——system of sewerage.
3. *Cemeteries.*
4. *Water Supply.*
    Private——use of wells——dangers——public——franchise, how obtained——terms——charges——use of public supply for domestic purposes——for fires——people who most need public water supply——municipal ownership of public water supply——compare Chicago, Buffalo, Manchester, (England), etc.

XIII. LIGHTING

1. *Public and Private.*
    Electric——gas——oil.
2. *Franchises.*
    How obtained——terms——cost.
3. *Municipal Ownership of Electric Lighting and Gas Plants.*

XIV. FINANCE

Assessment——taxation——revenue——expenditures——comparative cost of city governments of United States and foreign countries.

XV. BUSINESS

1. *Home Products.*
    Grain——live stock——timber——brick clays, etc.
2. *Sources and Cost of Power.*

3. Manufactures.
  Whiskey——beer——tools——barrel stuff——cars——iron goods——brick——quantity and value.
4. Wholesale Houses.
  Character——volume of business——territory supplied.
5. Retail Houses.
  Character——volume of business.
6. Banks, Insurance, Building and Loan Associations, etc.

XVI. DISTRIBUTION OF WEALTH

Individual wealth——how acquired——rich men——incomes——cost of living——expenditures for food——clothing——rent——luxuries, etc.

XVII. LABOR

Labor organizations——strikes——wages in different employments.

XVIII. PROFESSIONS

Divinity——law——medicine——teaching——journalism——engineering——cost of preparation for——incomes and salaries.

XIX. EDUCATION

1. Public Schools.
  Organization——courses of study——buildings——attendance——school population——teachers——salaries——cost——Indiana State Normal School and its relations to the city.
2. Private Schools.
  Rose Polytechnic Institute——Coates College——Commercial Colleges——teachers of music, dancing, etc.
3. Libraries.
  Size——circulation——character of books read——influence, etc.
4. The Press.
  Newspapers and periodicals published and circulated in the city——newspapers and periodicals published elsewhere and circulated in city.
5. Art Galleries.
  Exhibits and collections.
6. Literary, Art and Musical Societies and Clubs.
  Number——membership——character of work——lecture courses——concerts, etc.

XX. RELIGION

Churches——buildings——value of property——income——members——attendance——clergy——Sunday schools——missions——Young Men's Christian Association.

XXI. CHARITIES

Hospitals——dispensary——alms house——orphan asylums——institutions and societies——what is being done for the dependent classes——

for tramps——for criminals——for children——causes of poverty and crime——prevention and relief of poverty and crime.

XXII. RECREATION AND AMUSEMENTS

Hunting——fishing——boating——cycling, etc.——parks——athletic sports——races, etc.——public and private entertainments——plays, etc.——picnics——excursions——summer resorts——places of resort and amusement for laboring men——should they be open on Sunday?

XXIII. HOME LIFE

In families of different races——incomes——education——intelligence and social position——relations of husband and wife——divorces——relations of parents and children——relations of home life to idleness, vagrancy, vice and crime.

XXIV. SOCIAL ORGANIZATION

Groupings on basis of race, religion, politics, wealth, occupation, etc. ——standards of taste, intelligence and morals——secret, fraternal and other societies——peculiar customs and habits——peculiarities of language and speech.

XXV. REGULATION OF CONDUCT

Influence of education, religion, home training, public opinion, law—— prevalent motives of action.

XXVI. FAMOUS MEN AND WOMEN OF TERRE HAUTE

XXVII. PROBLEMS OF MUNICIPAL ECONOMY

Is the problem of city management essentially a political, or an economic and business problem?——relation of political parties to city government——city government in the United States, past and present ——the civic and municipal *renaissance*.

1. *Physical Problems.*

Cheap, rapid and safe transit——cheap and efficient lighting——pure, cheap and abundant water supply, universally distributed——ample and efficient sewerage——suppression of nuisances——the smoke nuisance ——the whistle nuisance——the manufacturing nuisance——the garbage and dust nuisance, etc.

2. *Political Problems.*

Purity of elections——honest administration of government——expert and competent officers and employes——enforcement of law.

3. *Social and Moral Problems.*

Prevention and suppression of drunkenness, poverty, vice and crime ——help for dependent and criminal classes——increase of intelligence and morality——improvement in the cleanliness and beauty of the city.

The outline is intended to be as nearly exhaustive as possible, and is as well adapted for university students as for lower grades. The collection of facts could be well done by pupils of the seventh and eighth grade, and with them would serve as an ever-accessible and unrivaled field for the study of home geography, and as a basis for the study of cities in general. In any grade the following advantages may be derived from such a plan: (1) The use of individual experience and observation. (2) The acquirement of direct, personal and first-hand knowledge. (3) The value in itself, as information, of such knowledge of facts and conditions which intimately concern the welfare and conduct of every student. (4) The increase of general intelligence. (5) The mental discipline obtained by the classification of such facts, and the discovery of their relation to each other and to other facts. (6) The conclusions which may be drawn from them by inductive reasoning. (7) A basis in experience for the study of other cities and countries. (8) Practical lessons in the science and art of civics and economics. The higher advantages will be attained in greater proportion as the grade of the student is more advanced.

At the top geography runs insensibly into history, sociology and political economy; in a word, into the new and comprehensive science of *demology*. It is scarcely worth while to try to discover the cleavage plane between them. This paper is offered as a contribution to the method of study along this plane.

X.—A SHORT HISTORY OF THE GREAT LAKES*

FRANK BURSLEY TAYLOR, F. G. S. A.

INTRODUCTION

In studying the geographical development of Indiana, while we are, of course, concerned chiefly with the area of the state itself, a thorough and comprehensive treatment of the subject would require the consideration of some contiguous areas many times larger. For Indiana has not developed in any sense as a separate unit, but rather as a fractional part of a very much larger area. One of the most important outside influences, especially in its bearing upon the development of Indiana's physiography, has been the existence of the valleys or basins of the Great Lakes, which lie toward the north. The direct influence of these basins has been comparatively small, but indirectly, through their effects upon the continental glaciers or ice-sheets, their influence has been quite important. This article, however, is designed to present a brief outline of the whole lake history, so that it will not be possible to dwell at much length upon its relations to Indiana.

THE PRESENT LAKES

The Laurentian, or Great Lakes of North America, form the largest system of fresh-water bodies in the world. No other, unless it be the unconnected group of lakes in Central Africa, bears any comparison with it. The Victoria Nyanza, which is the largest fresh-water lake known elsewhere, has an estimated area of between 25,000 and 30,000 square miles. The area of Lake Superior is 31,200 square miles, of Lake Huron 23,800 and of Lake Michigan 22,450. Lakes Erie and Ontario are considerably smaller, the surface of the former being only 9,960 square miles and the latter 7,240. The total area of the whole lake system with its connecting channels is 95,275 square miles.

The water-shed of a lake is the land surface which drains into it. Lakes Superior, Michigan and Huron each have water-sheds considerably less than twice the area of their own surfaces. That of Lake Erie is about two and a half times its water surface, and that of Lake Ontario three times. The mean annual rainfall of the St. Lawrence basin is about

---

* This article has been revised in the light of the latest discoveries, especially for this re-issue. The text has been largely re-written and the maps are all newly drawn.

thirty-one inches, and the mean depth of water evaporated annually from the surface of the lakes is something between twenty and thirty inches. The amount of precipitation on the water surface is therefore nearly compensated by the amount evaporated from the same area, so that the volume of the outflow of the lakes is almost exactly equal to the rainfall on their water-sheds, less all the evaporation that takes place from the land and all the rivers and small lakes before the water reaches the Great Lake basins.

Lake Superior has an elevation above sea level of 602 feet; Lakes Huron and Michigan 581; Lake Erie 573, and Lake Ontario 247. The greatest depth of Lake Superior is 1,008 feet; of Lake Michigan 870 feet; of Lake Huron 730 feet; of Lake Ontario 738 feet, and of Lake Erie 210 feet. It follows that four of the basins reach depths below sea level. Lake Superior reaches 406 below; Lake Huron 149; Lake Michigan 289, and Lake Ontario 491. The mean depths of the lakes, however, is considerably less, the greatest being Lake Superior at 475 feet, while Lake Michigan is about 325, Lake Ontario 300, Lake Huron 250 and Lake Erie only 70 feet.

The volume of discharge of the several outlets is as follows: St. Mary's river, outlet of Lake Superior, 86,000 cubic feet per second; St. Clair river, outlet of Lake Huron, 235,000; Niagara river, outlet of Lake Erie, 265,000; St. Lawrence river, outlet of Lake Ontario, 300,000. The volume of water contained in all the lakes taken together is about 6,000 cubic miles, of which Lake Superior contains a little less than half. This would keep Niagara Falls running at its present rate for about one hundred years.*

General Geological Relations

In their general geological relations the Great Lakes are only a part of a much larger system of basins which reaches, in fact, clear across the continent from the Atlantic coast to the shores of the Arctic Ocean. Lake Winnipeg, Lake Athabasca, Great Bear Lake and Great Slave Lake all belong to the same chain and some of them are very large. There are several other depressions which have the same relation to the geological structure, and are of the same age, but which are now occupied by arms of the sea. Of these the Gulf of St. Lawrence and the Gulf of Maine are the largest. Most of the lake basins are excavated out of what are called the Paleozoic rocks and mostly along the line of contact between these and the crystalline rocks, granites, gneisses, etc., of an older or Archaean age. The lakes mostly lie where the edges of the Paleozoic strata are slightly upturned, and were in consequence weak and more easily eroded away. This is the case with the Ontario, Erie, Huron, Georgian

---
*These statistics are taken from "The Lakes of North America," by Professor I. C. Russell, of the University of Michigan.

Bay and Michigan basins. Lake Superior, however, is probably an exception. Counting the height of its coast lands, its basin is much the deepest of all. Although it has crystalline rocks near part of its south side as well as on the north, east and west, there is, however, some evidence that its basin was once occupied, at least in large part, by softer rocks of later ages. In the Ontario and Erie basins the edges of the strata project slightly upward towards the north; in the Huron and Georgian Bay basins towards the east, northeast and north, and in the Michigan basin towards the west and northwest.

Theories of Origin

A number of theories have been advanced to account for the origin of the Great Lakes. One of the earliest was that of the famous French traveler, C. F. Volney, who came to this country in 1783 to study its soil and climate. He was much impressed with the greath depth of Lake Ontario, especially near its south shore, and with the high, bold cliffs of limestone that stand only a few miles back. Volney concluded that Lake Ontario is the crater of an "extinguished volcano," and that it was made by volcanic explosions which blew away the rocks that once filled the basin. But Volney was only keeping in line with the contemporary idea, for DeLuc, who was one of the most eminent geologists, held, twenty years later, that the widely scattered or "erratic" boulders which cover the plains of northern Switzerland had been fired as bombs out of volcanoes. It was left for Louis Agassiz and his contemporaries, about sixty years ago, to prove that these boulders were carried out from the Alps and spread over the plain by slowly moving masses of ice—by great glaciers now extinct.

The lake basins have also been supposed by some to be rock basins or hollows made merely by the crumpling and folding of the rocks of the earth's surface. We shall see later that none, except possibly Lake Superior, in a qualified sense, are of this character, although earth movements appear to have had much to do in other ways with their formation.

Probably the first steps toward a scientific inquiry into the origin of the Great Lakes were those of the early followers of Agassiz. He first pointed out the fact that the whole region of the Northern States and the adjacent parts of Canada had been completely covered by a mighty glacier. The drift which overspreads these regions was appealed to by some as evidence of the glacier's power to scoop out solid rocks.

But the fanciful guesses of Volney and the theories of the early glacialists have gradually given place to the results of methodical exploration, so that to-day the making of the basins of the Great Lakes is attributed chiefly to four distinct causes; viz., to the wearing, eroding action of streams; to the uplifting and tilting of the land; to the obstructing action of the glacial drift and to the wearing-down or abrading action of the glacier itself. All these causes have played their part. The lake basins were not gouged out entire by the ice-sheet, although it undoubtedly widened them to some extent by tearing away the weaker ledges of their sides, especially where the ice mass moved against the rocks with full force. But the ice-sheet had little or no tendency to deepen them. They are mainly old river valleys uplifted and tilted by movements of the solid earth and choked up and obstructed here and there by a glacial drift.

PRE-GLACIAL AGES (Pre-Pleistocene)

THE PRE-GLACIAL CYCLE OF EROSION

The northeastern quarter of the continent, including Indiana and probably all of the Great Lake region, was raised up out of the sea some time not long (speaking in a geological sense) after the last of the coal beds of Pennsylvania had been deposited. It is not possible to say when that was, even approximately, in years, but all authorities seem to agree that it was at least several millions of years ago. The thing for us to note especially here, is the very great age of this part of the continent as a land surface. Through all the long ages since the Appalachian mountains were uplifted the region of Indiana and the Great Lakes, and probably the whole of Canada lying north of the lakes and the St. Lawrence river, have been a land surface, with sunshine, rain, wind, frost, chemical solution, creeks, rivers, the waves of lakes and the sea, vegetation and animal life constantly at work upon it, sculpturing and wearing down its surface. Mild climatic conditions had prevailed, and at the end of this long cycle of erosion, the whole surface of the land was deeply decayed, soft, almost rotten. The rivers had gone on for thousands and thousands of years deepening and widening their valleys and no disturbing element had intervened to seriously interrupt or modify their work. The decay of the surface was probably not so deep as that of the southern Appalachians, as may be seen, for instance, at Asheville, North Carolina, where uninterrupted decay has gone on for a still longer time. But it was certainly not very different from that. It probably resembled that part of Wisconsin which was not invaded by the ice of the glacial period, and the unglaciated area in Southern Indiana. In such regions, besides deeply decayed rock and soil, there are cliffs, "chimneys," "sugar-loaf" rocks and other very old and frail relics of subaerial erosion. As a product of the long decay, a vast amount of loose material was left covering the whole surface of the country. In this way, chiefly, was prepared the enormous quantity of debris which was afterwards crushed and ground up by the ice-sheet, transported in large quantities by it hundreds of miles, and finally spread over the surface of regions far away.

When the first ice-sheet began to gather upon the highlands north of the St. Lawrence river, the surface of the lake region presented a very different appearance from what we see to-day. There were probably, then, no great lakes like those of the present time in this region, but only wide open valleys instead. Or, if the lakes were already in existence, they had probably only very recently been produced by an uplift of the land which had tilted the old valleys in such a way as to cause them to hold water.

When the first ice came, the cutting down of the land by the streams had been far developed. The greater rivers had wide, open valleys; the lesser rivers flowed in valleys of comparatively great depth but moderate width, while the little creeks and brooks had each its own deep, narrow, steep-sided gorge. This was especially true of the country bordering the lakes on the south, as in Western New York and Ohio, while in the hard, crystalline rocks of the north the topography developed was still more rough, as in the region east of Georgian Bay.

The proof of this condition is found in the discovery of many valleys deeply buried under the drift, and in occasional remnants of the old decayed surface itself. Buried valleys and gorges are common throughout the whole drift-covered area as far as known. One now filled with glacial drift, cuts across the neck of land between Lake Erie and the west end of Lake Ontario, as though Niagara Falls had sawed its way through that same region before. In Western New York, Pennsylvania and Ohio many buried valleys have been found of rivers of considerable size which formerly flowed northward to Lake Erie, opposite to their present courses. The upper Ohio is one of these, which had its valley to the north filled with drift, and was in consequence turned into a lake which overflowed to the west and finally cut down the obstructing barrier on that side and found a course to the Mississippi. Indeed, we owe it to the last glacier that the Great Lakes are what they are to-day—brim full and standing nearly 600 feet above the sea. The tilting up of the land, alone, would have made the Lakes in the first place. But Niagara Falls would have speedily cut back to Lake Erie and drained it off, and probably also lowered the level of Lakes Michigan and Huron, while the St. Mary's river would soon have brought Lake Superior down also. Ontario and the three upper Lakes would have continued to exist, much reduced in magnitude, but Erie would probably have become extinct. Thus the glaciers, and the last one in particular, have acted as restoratives by choking up the narrow parts of the outlet valleys, and so raising the level of the waters and compelling the rivers to find new courses and begin their cutting over again.

The state of Indiana was a part of the ancient land surface, and suffered the prolonged decay and erosion along with the rest. In the unglaciated part of the state south of Indianapolis, the old eroded surface may be seen to-day, and in the western part of the state there are some places that show it only slightly modified by glacial action. There are probably many buried valleys in the northern and northeastern parts of the state where the drift is very deep, but none have yet been found equal to some of those in neighboring states. A boring near the northern state line passed down through a great depth of drift and reached the old rock surface at a level slightly below the present surface of Lake Erie. Too little, however, is known as yet, concerning the rock surface beneath the drift to enable us to say much about it. We do not know what its drainage system was. Although Indiana fronts on Lake Michigan along a space of about forty miles at its northwest corner, the present drainage of that part of the state is determined by the configuration of the drift surface and only a very narrow strip, about ten miles wide, along the shore drains into the lake. It is possible that in pre-glacial times the drainage of Indiana was more intimately related to the lake valleys. The relations of the present time, however, afford no indication whatever of the earlier conditions of drainage. A small area along the northern line of the state now drains to Lake Michigan through the St. Joseph river of Michigan, and another small part drains eastward to Lake Erie through the Maumee river. But the arrangement of both these systems appears to be entirely dependent upon the present drift topography, so that the earlier drainage of both areas may have been entirely different.

THE LAKES IN THE GLACIAL OR PLEISTOCENE PERIOD
The Lake Basins and the Ice-Sheets

The glacial period, properly speaking, comprises the whole period of time from the first advance of ice as an ice-sheet down to its last and final disappearance. The question whether the period, as a whole, comprised two or more distinct epochs of glaciation with warm epochs between, or only one with minor variations, has been very much in controversy for some years. But the facts have constantly gathered in support of those who favor diversity, rather than of those who favor unity, and it now seems to be well established that there have been at least two or three, or possibly even four or five, distinct epochs in the glacial period, each with its own great ice-sheet.

The growth of the ice-sheets was extremely slow. For some as yet unexplained reason the climate gradually grew colder, and the winter's snows in the regions surrounding Hudson Bay began to exceed the summer's melting and evaporation, and as this went on year after year and century after century, a small ice-sheet formed on the high plateau of the Laurentide mountains north of Quebec, and gradually spread away over all the surrounding country. Other sheets like it were started in other high parts of the north, and all of them finally blended together as one grand glacier covering the whole northeastern quarter of the continent. When the maximum extent of the ice was attained, the whole region of the Great Lakes was deeply buried under it, and the front of the ice reached southward to Nantucket, Long and Staten Islands; extended across New Jersey and Pennsylvania in a zigzag line to the southwest corner of New York; then across the northwestern corner of Pennsylvania and Central Ohio to a point near Cincinnati, where it pushed over a few miles into Kentucky; then back into Southern Central Indiana; across Southern Illinois, Central Missouri, Eastern Kansas and Nebraska; and finally across the Dakotas and Montana to British America. [See map, page 33.] Then, for some equally unexplained reason, the great glacier halted, and finally withdrew again northward to the obscure regions whence it came, and its retreat was in the same order and as gradual as its advance had been. The advance and recession in each case was accompanied by many minor oscillations, with re-advances apparently periodic, as though the ice front had gone two steps forward and one back in advancing, and one forward and two back in retreating.

This is substantially the story of each of the separate ice-sheets. But each successive one varied a little from its predecessor in the limits of its advance, and so did not stop at the same line. In the west, each one fell short in most places, of the mark reached by its predecessor, while in the east, it appears probable that the last one in some places overreached all its predecessors. The ice of each sheet carried boulders, gravel, sand and clay frozen fast in its lower layers, and it was continually dropping some part of what it had and picking up more. Enormous quantities of this material lodged in the depressions in the rock surface beneath. But part of it was carried forward to the front where it was dropped at the melting edge. Where the ice front stood a long time in one place, a ridge of debris accumulated and was left on the final withdrawal of the ice as a terminal moraine.

There is a complex series of these moraines extending back northward from the extreme front of the last ice-sheet. Their chief interest to us in connection with the lake history is that they show clearly how the ice-sheet was related to the lake basins at successive stages of retreat. As the ice came down from the north and northeast over the uneven surface of the country, it had more or less pronounced currents of flow. It naturally moved faster and farther forward in the main valleys where it met the least resistance, and lagged behind in the higher, hilly regions. When the ice began to rise above the southern rims of the basins it spread away over the adjacent plains in great sheets that fitted themselves to the wide, flat valleys with marvelous exactness, and even felt the influence of the lesser topographic features.

At this stage the ice-front had a lobate form, each lake and great bay marking the place of a southward projecting blunt arm or lobe of ice. The ice-lobes spreading from the several basins finally blended together, and when the whole sheet had reached its maximum extent the lake basins had lost much of their influence upon the flow, and the final front line fitted itself to the local topography almost as though the lake basins had not existed. At its climax the ice flowed over the great lakes as a river flows over a hole in the bottom of its bed. When the ice-front had retreated nearly to the lakes it again became segregated into distinct lobes with sharp, re-entrant angles between. The lands between the lakes were the first to be left bare, and the ice lingered last upon the north and northeast sides of the basins.

Three glacier lobes, corresponding to as many basins toward the north, entered Indiana: the Erie (with which was combined the lobe from Lake Huron) covered the eastern part of the state, and the Saginaw and Michigan lobes combined to cover the northern and western parts. The combined effect of the extent, relative strength, oscillations, conflicts and the relative positions of these lobes was the prime factor in shaping the topography of the northern half of the state. With the probable exception of the Wabash below Attica, every stream in this area had its course determined or largely modified by the features of the drift, and especially by the moraines. If the lake basins had been absent or differently located, or if the ice had advanced from a different direction, the drainage systems and the general arrangement of the physical features of this part of Indiana would have been entirely different.

The Earlier Ice-Dammed Lakes of the Glacial Recession

In the region of the Great Lakes the front of the ice-sheet retreated in a general north-northeasterly direction, in some places more nearly east and in others more nearly north, according to local influences. The present outlets of all the Lakes, except Lake Huron, are from their eastern or northern sides. Hence, as the ice-front moved slowly back, it at length withdrew to a position behind the drainage-divides south of the Lakes and uncovered a little of the watershed of some of the lake basins. At these places lakes were formed by the collection of water from rain and from the melting ice, and in each case the surface of the lake so formed was held up to the lowest level of the divide recently uncovered and this point became its outlet. To the north and northeast the great glacier still stretched away for hundreds of miles as a continental plain of ice. Its deep, solid mass performed the same service, temporarily, that the land does to-day: it acted as a great dam, and effectually prevented the escape of the water in that direction.

The exploration of the several lake basins has revealed a large number of old shore lines and outlets, all now abandoned. Enough has already been learned to show that nearly all of them were made by ice-dammed lakes during the glacial recession. In the beginning these glacial lakes

Figures 1 and 2 represent the first and fourth stages of the earlier glacial lakes. The exact position of the ice-front all along the line is not yet accurately known for any particular stage. In Figure 1 the map of the Erie lobe and Lake Maumee is based on accurate information, but the contemporary positions of the other lobes are known only approximately. They may have stood a little farther forward or a little farther back than here represented. In Figure 2 the features are all accurately known except the precise positions of the Erie and Michigan lobe-fronts.

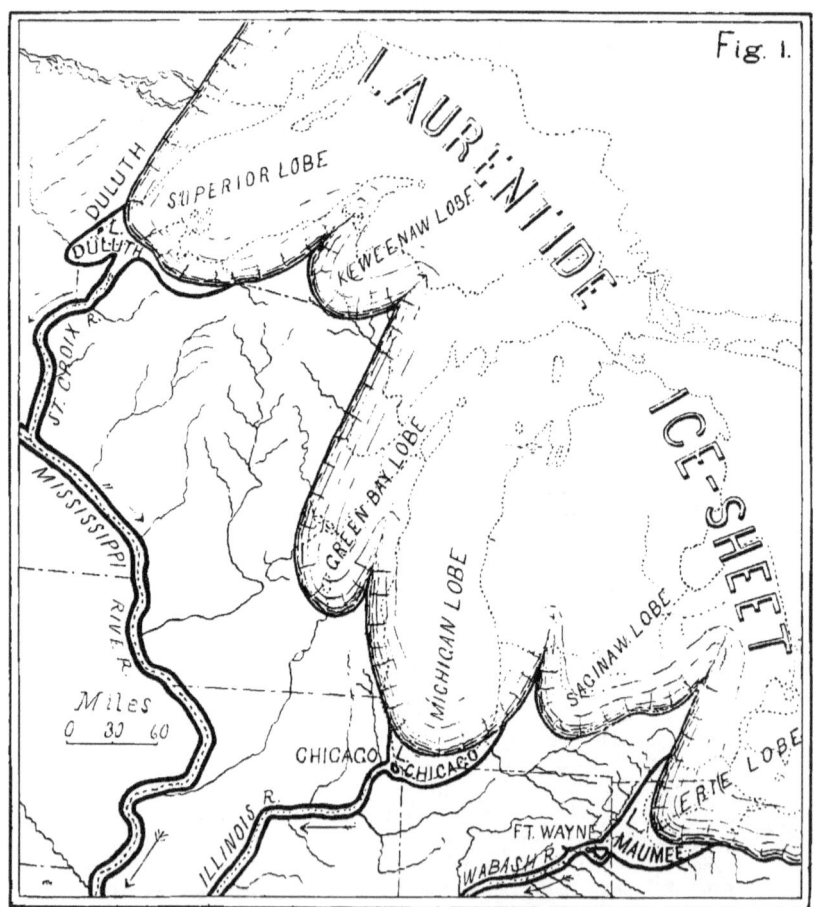

Figures 1 and 2.—TWO STAGES OF THE EARLIER GLACIAL LAKES.

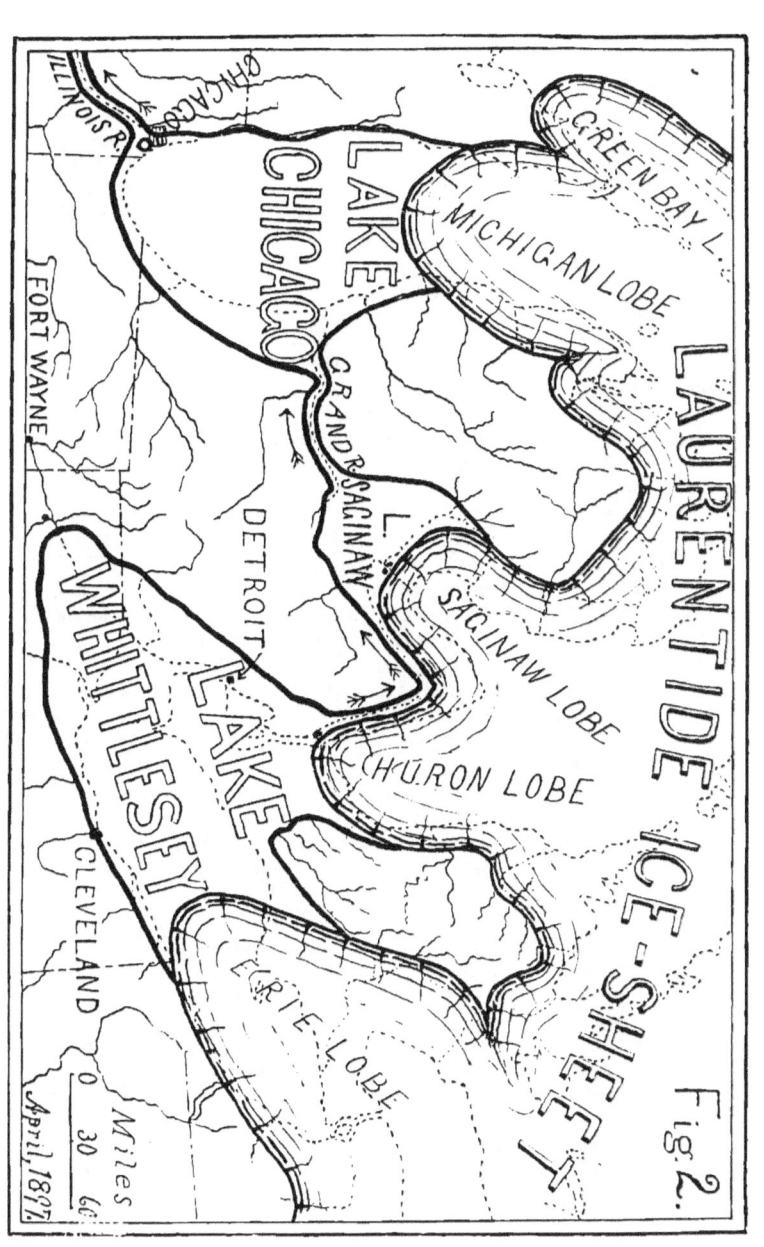

were small, and each one had its own outlet and was entirely independent of the others. But with the progressing recession of the ice they grew larger and larger until the dividing lands between them were left uncovered. Then contiguous pairs of lakes united, and always at the level of the lower one. In this way the lakes kept combining and finding new outlets at lower levels as the recession went on. The stability of each lake depended for the time being upon the position of the ice-front and its relation to the higher parts of the adjacent uncovered land surface. This makes the relation of the lakes to each other intimately dependent upon the direction of the retreat. It the general retreat had been directly north over the whole area the history of the glacial lakes would have been entirely different from that of a retreat directly east or northeast.

The lakes that were formed in front of the ice-sheet during its retreat across the Great Lake region were very numerous and their relations and changes very complex. A detailed account of their history would require much more space than can be taken here. Figures 1 and 2 will help to give a general idea of the changes. The history of the glacial lakes of the Erie and Huron basins are most fully known and will serve best as an illustration of the kind of changes that took place.

One of the earliest of the ice-dammed lakes was Glacial Lake Maumee Fig. 1), formed in the western part of the Maumee valley when the ice-front had retreated to the east from Fort Wayne, which is on the lowest point of the western rim of the watershed of Lake Erie. This lake covered part of Ohio and Indiana and had its outlet westward through Fort Wayne to the Wabash river at Huntington. It had its most permanent form when the ice-front extended in a great curve convex to the west from Findlay, through Defiance to Adrian. Professor C. R. Dryer has described this lake in another chapter, and the reader is referred to that for fuller details. But Lake Maumee was only a beginning. For as the glacier continued its retreat, the lake grew larger and larger, and would have continued to expand indefinitely, except for the fact that a lower notch in the lake rim than that at Fort Wayne was at length uncovered. This first lower passage was at Imlay, Michigan, due north of Detroit and about west of Port Huron. The discharge in this direction lowered the lake permanently twenty-five or thirty feet. The water that went out this way, flowed southwest across Michigan finally reaching the Mississippi.

As the retreat went on, the ice in the Erie basin finally separated from that which was coming southward from Lake Huron; one lobe retreating northeastward down Lake Erie and the other northward up the St. Clair valley towards Lake Huron. But the two separate ice-dams continued to hold the lake as before.

With further retreat, the Huron lobe finally uncovered another passage lower than the one at Imlay. This new outlet was on the "thumb" well out towards its end, about north-northwest from Port Huron. (Fig. 2.) To this new notch the level of the waters dropped again. By this time the retreating Saginaw glacier also held a lake in front of it, so that at this stage there were three lakes forming a chain; one in the Erie basin, one in the Saginaw basin and one in the Lake Michigan basin. These three lakes have been named Lake Whittlesey, Lake Saginaw and Lake Chicago respectively. The water flowed westward through the chain, and finally through the southern part of the city of Chicago and down the Des Plaines and Illinois rivers to the Mississippi. At the next step of retreat the "thumb" of Michigan was left free of ice, and Lake Whittlesey fell and blended with Lake Saginaw. This new combination is called Lake Warren. The chain of three lakes was thus reduced to two, Lake Warren and Lake Chicago, and this arrangement lasted for a considerable time.

At its maximum extent Lake Warren covered the south half of Lake Huron, including Saginaw Bay, the whole of Lake Erie and the low ground between it and Lake Huron, extended eastward to within twenty or thirty miles of Syracuse, N. Y. and probably covered some of the western end of Lake Ontario.

In the meantime the ice was retreating in the east as well as in the west, and uncovering lower ground in that direction. Finally, in Central New York an open passage was left along the ice-front to the Mohawk valley at a lower level than the outlet to Lake Chicago. Then the discharge of Lake Warren turned eastward, and the level of the waters fell so as to uncover the land between Lake Huron and Lake Erie. Across this the discharge of the upper lakes began to flow, and with the continued falling at the east, the water in the Ontario basin soon dropped below the level of Lake Erie, and then Niagara river and the Falls came into existence. About the same time, or soon after the fall of Lake Warren, the ice had so far withdrawn from the northern basins as to allow the waters of Lake Superior, Michigan and Huron to unite as one lake with its outlet through the St. Clair river to Lake Erie.

THE ICE-DAMS

If we wish to obtain a realizing sense of the massiveness of the ice-dams that held these lakes up, we may readily do so by taking account of some of the facts which have been revealed by the study of the drift. At Defiance, the ice-front stood in about sixty feet of water; at Saginaw in about 150 feet, and at Toledo, Detroit and Port Huron in about 200 feet. And yet, the ice-lobe in each case kept its place and fitted itself to the form of the valley, as revealed by the flat, low, water-laid moraines at these places, apparently as perfectly as though no standing water had been present. This shows clearly how massive and solid the ice must have been. It did not break up much and float away as icebergs, but was able to withstand wind and wave and its own bouyancy or tendency to float in the water. Its front in the water must have been undercut by the waves and reduced to a great perpendicular or perhaps even overhanging cliff of ice; but it kept its place, and built its moraine almost exactly where it would have built it if no water had been present.

These lobes must have been at least 300 to 400 feet thick close to their edges in the lakes, and they must have been solid and compact and comparatively free from cracks and crevices. This shows that their forward motion must have been exceedingly slow, for rapidly moving glaciers are always riven by many deep cracks and crevasses, which cause them to break up and float away easily when they enter deep water.

By tracing the single terminal moraine that was made at the edge of the ice when its front points stood at Port Huron and Saginaw, (during the time of Lake Whittlesey), the immense proportions of the ice-dam have been disclosed (Fig. 2). The apexes at the two places mentioned stood on ground only a few feet above the present level of Lake Huron. But, at the same time, the ice-front rested on the highlands 130 miles north of Saginaw, and also on those in Canada 180 miles northeast of Port Huron, at an altitude of over 1,000 feet above the present lake level. The cross-section of the glacier between these two highlands (about 200 miles apart) probably had a slightly arching surface, like the ice-cap of Greenland, so that the depth of the ice in the middle of the lake measured upward from the present lake surface, must have been somewhere near 1,500 feet.

When the ice-sheet covered most of Indiana and crossed the Ohio river, the depth over central Lake Huron must have been much greater, probably two or three times as great. At that time the ice was at least 500 or 600 feet deep over the present site of Terre Haute, and nearly as deep over that of Indianapolis, and it thickened gradually northward. If an observer could have stood on one of the hills in Brown county at that time he would have seen to the east of him the great wall of the ice-front extending south towards Kentucky, while towards the west it would have been seen in the distance stretching away towards the southwest. For hundreds of miles to the east and west, and for 2,000 miles or more to the north, the glaring, white desert of snow-covered ice, like that seen in the interior of Greenland by Nansen and Peary, would have appeared, stretching away out of sight with not a thing under the sun to relieve its cold monotony. It is hard to think of Indiana and her neighboring sister states as being clothed in such a shroud-like mantle as this. But it was in large part this same ice-sheet, coming perhaps four or five times in succession, that covered the state with the inexhaustible soil of the drift, and made Indiana the fertile agricultural state that she is to day.

In Figure 3 both positions of the ice front are located with only approximate accuracy. Each is placed in the last position in which it could have held the lake in front of it. The position shown for the Ontario lobe is earlier in time than that of the Ottawa, for Lake Iroquois was drained off before Lake Algonquin.

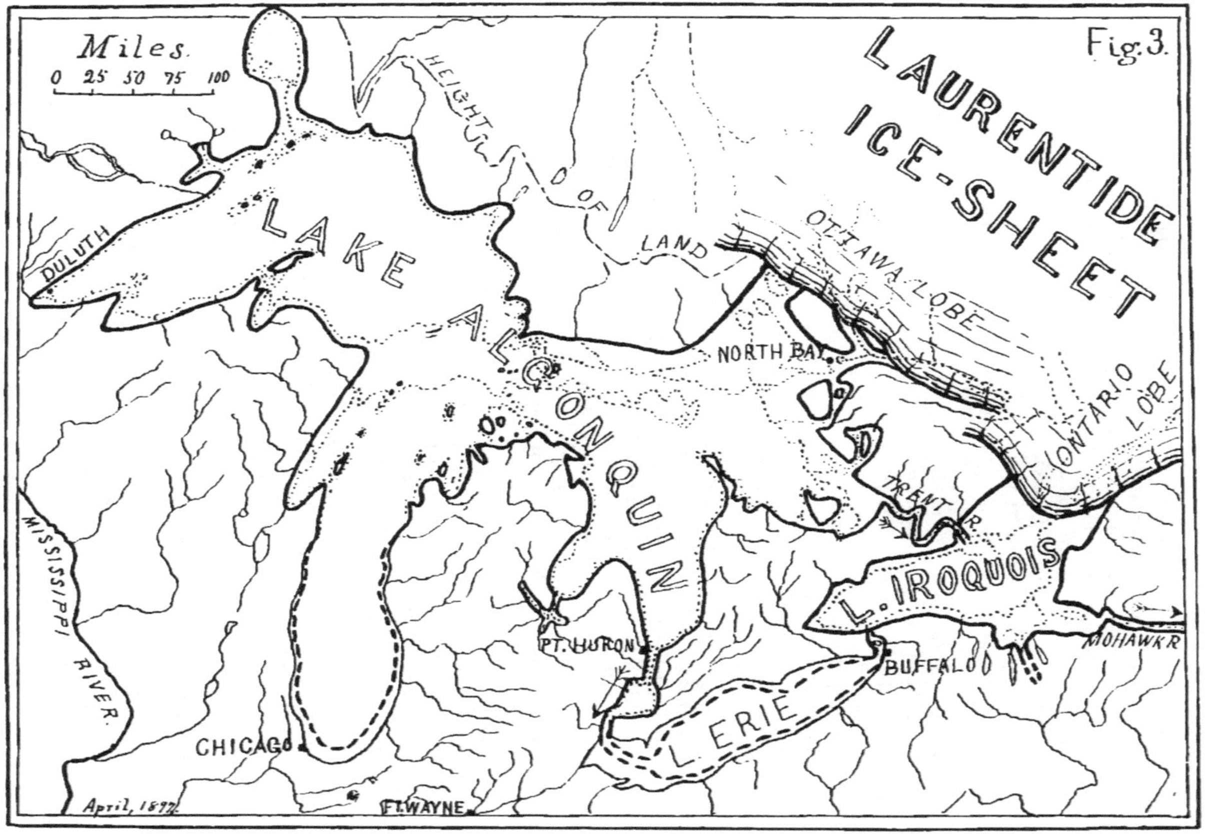

Figure 3.—GLACIAL LAKES ALGONQUIN AND IROQUOIS.

Glacial Lakes Algonquin and Iroquois

After the fall of Lake Warren, Lake Erie became independent, and only two large glacial lakes remained in the Great Lake area. (Fig. 3.) One filled the three upper basins and is known as Lake Algonquin. Its outlet was south to Lake Erie; for the ice-sheets still covered the lands to the east and northeast of Georgian Bay, where the only other lower ways of escape were situated, as the land stood at that time. The Chicago outlet was then apparently about 100 feet above the lake level, and so did not serve as an outlet for Lake Algonquin. Finally, the retreating ice uncovered Balsam Lake at the head of the Trent valley in Ontario, east of the south end of Georgian Bay, and the discharge of Lake Algonquin shifted to that place. The erosion effects of the great outlet river are quite plain along the course of the Trent valley. The modern Trent river is a comparatively small stream. Probably the head of this outlet was not much below the level of the St. Clair outlet at that time, so that the change produced only a slight lowering of the lake.

About fifty miles northeast of the north end of Georgian Bay lies Lake Nipissing in an east-and-west trough which leads through the highlands to the Ottawa valley. Lake Nipissing, itself discharges, westward through French river to Georgian Bay. But it is only three miles across a low, swampy divide at the head of the lake to Trout Lake which discharges eastward through the Mattawa river into the Ottawa. This old divide, called the Nipissing Pass, is somewhat less than 100 feet above Georgian Bay, and the town of North Bay is built upon the west end of it on the shore of Lake Nipissing. To the eastward for about 100 miles the Mattawa and Ottawa valleys extend as a narrow trough, 700 to 800 feet deep. As things were then, this trough was much the lowest opening in the rim of Lake Algonquin; but, so long as it was filled with ice, the lake kept its level. Thus, a relatively small ice-dam was able to maintain a very large lake. This dam continued to hold the water in place until the ice filled only twenty-five or thirty miles of the lower, narrow part of the Ottawa valley. If no other changes had interfered the lake would have dropped its level 500 feet when the dam broke. But before the dam gave way there were upheavals of the land which tilted up the whole region northeast of the lakes, so that when the break finally came the lake dropped much less than 500 feet, and it probably fell rather slowly.

These same upheavals produced another important effect. They shifted the discharge of Lake Algonquin back again from the Trent valley to the St. Clair river, and this restoration took place a considerable time before the Ottawa ice-dam broke. At its maximum, Lake Algonquin was considerably larger than the present combined areas of Lakes Superior, Michigan and Huron, including Georgian Bay.

The water that gathered in the Ontario basin was held up by the ice-sheet which formed a dam across the St. Lawrence valley northeast of the lake. This glacial lake is known as Lake Iroquois. Its outlet was through the Mohawk valley, and at its greatest extent it was considerably larger than present Lake Ontario. After Lake Algonquin had begun its career, it was some time before the eastern water fell to the level of Lake Iroquois. Lake Algonquin endured also for a considerable time after Lake Iroquois had been drained off; so that the former was the longer-lived of the two and was, in fact, the longest-lived of all the glacial lakes within the St. Lawrence basin.

The reality of the great ice-dams that held up the larger lakes is no longer to be doubted. For within the last two or three years the beds made by the temporary rivers that drained them off when the dams gradually gave way, have been found and partially explored. Those of falling Lake Warren are in Western New York, those of Lake Iroquois on the northeastern flanks of the Adirondack mountains, and those of Lake Algonquin, less fully explored, are in the Mattawa and Ottawa valleys.

The Lake Beaches

One of the most remarkable things about the old shore lines of the Great Lake region is the fact that they are not horizontal when compared with present water levels. The beaches at the western end of the Erie basin, and the Algonquin beach in the east half of the Superior basin are substantially horizontal. But all the rest are more or less inclined upward in a northeasterly direction. The inclination is not the same in different beaches, being generally greatest in the older and higher ones; and it varies considerably in the same beach in different places. That all the beaches were horizontal when they were made seems certain. It follows that their present departure from that attitude is a measure of the amount of upheaval of the land since they were made. The older, higher beaches record the net result of many changes. But the lower, younger beaches record only such changes as occurred after they were made. Hence, theoretically, the deformation of the latter ought to be generally simpler and show fewer irregularities.

The Algonquin beach rises from twenty-five feet above the lake at Port Huron to 635 feet above it near North Bay, Ontario, and it is a little over 400 feet above Lake Superior in nearly the whole of that basin. It is 205 feet above the lake at Mackinac Island, but descends southward and passes under Lake Michigan probably about 100 feet at Chicago. The Iroquois beach also rises toward the northeast in the Ontario basin.

THE LAKES IN THE POST-GLACIAL OR POST PLEISTOCENE PERIOD

The Nipissing Great Lakes

We come now to a part of the lake history which has scarcely any connection with Indiana, but which is of great interest to geologists and geographers generally. This interest arises partly from the fact that this epoch of the lake history is so recent, certainly mostly within the period of human occupation of America, and partly from its intimate relation to Niagara Falls and its gorge and the bearing which these have upon the date of certain great changes which have taken place in very recent geological time in the northeastern part of the continent.

When the water went out of Lake Algonquin, the glacial history of the Great Lakes came to an end. The water in the three upper basins then fell to the level of the Nipissing pass and it became their permanent outlet. (Fig. 4.) This arrangement lasted for a relatively long period of time, for the beach which the waves of that time made around all three of the basins is the strongest and most heavily developed of any in the Great Lake area. On account of its association with the Nipissing pass, it is called the Nipissing beach. Although it is now found at heights of more than 100 feet above the present lakes in the extreme northeast, the connecting channels between the three lakes of that time were only a little wider or deeper than those of to-day. In Mackinac Straits the Nipissing beach is about forty-five feet above Lake Huron, and the strait was formerly about a mile wider. At Sault Ste. Marie the beach is about fifty feet above Lake Superior. The St. Mary's river of that time was about a mile and a-half wider than now, and was more like a strait than a river. The Nipissing beach was, therefore, the shore line of three great lakes which were almost as distinct, and had nearly the same relations to each other as the corresponding three lakes of today. So they are called the Nipissing Great Lakes.

Many evidences attest the long duration of the lakes at the level of the Nipissing beach. On a shore which is comparatively new it is a common thing to find bays almost cut off from the main lake by long slender gravel bars or spits which have been built out across their mouths by the waves. There are fine examples of this type on the present shore of Lake Erie about the west end, and on the east and south shores of Lake Ontario. But when a shoe line is old, like the Nipissing beach, such bays are either filled up entirely or are cut off by wide sand or gravel plains and turned into separate lakes. There are a large number of small lakes on the coasts of the present upper Great Lakes which were

---

Figure 4. The Nipissing beach is confined to the basins of Lakes Superior, Michigan and Huron as shown by the heavy line. This beach has been explored on nearly all the shores. From North Bay eastward forty miles to the Ottawa river, the scoured bed of the former outlet stream is very well marked. The shaded portion of the map represents the contemporary area of the sea. It entered the Ontario basin, but dipped under the present lake along the south side and the west end. The limits of the marine area are indicated with only approximate accuracy.

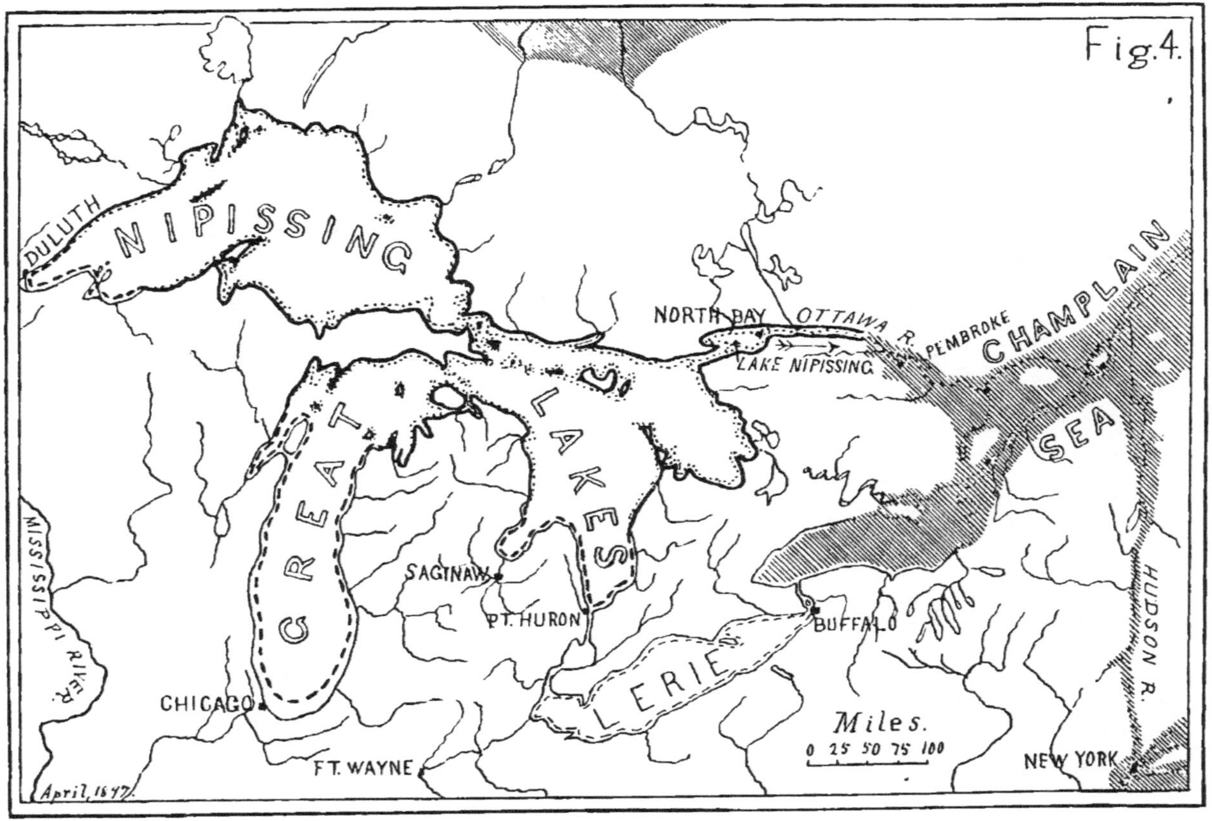

Figure 4.—THE NIPISSING GREAT LAKES AND THE CHAMPLAIN SEA.

produced in this way by the lodging of the shore-drift of the Nipissing beach.

In the Mattawa valley the work done by the outlet river shows that it flowed for a long time.

Like the older beaches above, the Nipissing beach is not now horizontal, but is tilted up at the northeast. The latter, however, is quite remarkable from the fact that although it is tilted, it shows no measureable irregularities in its plane. In this it differs in a marked way from the other beaches. From the northwest side of Lake Superior to the south side of Georgian Bay, and from the north end of Green Bay to the northeast corner of Lake Superior and the north side of Lake Huron the plane in which this beach lies appears to be perfectly uniform.

The direction of maximum rise is about north twenty-seven degrees east, and the rate of rise is nearly seven inches per mile. Thus, the Nipissing beach slopes downward toward the south-southwest and passes under the water of each of the present lakes. The depth at which it passes under the present lake level is estimated to be about forty feet at Port Huron, 100 feet at Chicago and twenty-five feet at Duluth.

The uplift which tilted the Nipissing plane up at the northeast, raised the outlet at North Bay, and would have raised the level of all three lakes correspondingly, had not another outlet been found. Soon after the tilting of the land began, the head of the St. Clair river became lower than North Bay and the outflow therefore turned to it. The arrangement of the lakes then inaugurated by this change has continued, apparently without interruption, to the present day.

A wide region lying east and northeast of the upper lakes has also had an eventful history in recent times. Its close connection with the lakes makes a consideration of some events of its history indispensable to a full understanding of the lake history.

The Champlain Submergence and Uplift

It has long been known that the northeastern part of the continent has been very recently uplifted out of the sea. Fossil shells of marine species, bones of whales and seals and marine fish have been found in various places, especially in the valleys of Lake Champlain and the St. Lawrence river, in deposits which were uplifted at that time. The amount of marine submergence was only a few feet at New York and Boston, but increased northward to over 300 feet in Maine and New Brunswick, over 400 at the northern line of New York and over 500 at Montreal and about 600 at points farther northeast. Marine fossils have been found as far up the St. Lawrence as Brockville, nearly to Lake Ontario, and up the Ottawa river as far as Lake Coulonge, a little below Pembroke. The bones of a whale were found in a gravel bed near Smith's Falls, Ontario, many years ago 440 feet above the sea, and the top of the beach in which they were buried is about thirty feet higher. The sea at that time extended through a strait into Lake Ontario, and up the valley of the Ottawa a considerable distance above Pembroke. (Fig. 4, shaded area.) The character of the fossils and also the very youthful condition of all the river beds below the old marine level, prove the recentness of the uplift that raised the fossiliferous Champlain beds out of the sea.

Turning our attention now to the upper lakes, we find that the same characters prevail below the Nipissing beach, except that there is no evidence of marine life. The very youthful river beds and shore lines below the Nipissing beach are quite marked, in contrast with the older character of the forms above, corresponding closely in this respect with the evidence of newness in the area of the Champlain Sea to the east. These two great areas, the marine on the east and the fresh-water on the west, lie close together, almost interlocking, so that it seems certain that it was one and the same uplift that affected both. When the Champlain beds were uplifted out of the sea, that same movement uplifted and tilted the area of the Great Lakes. The evidence points plainly to the conclusion that it was this uplift that tilted the Nipissing beach and shifted the outlet of the upper lakes from the Nipissing pass at North Bay to the St. Clair river at Port Huron.

Falls of Niagara

The clock by which we can time these events in a roughly approximate way is the cataract of Niagara. Lake Erie furnishes on the average about one-ninth of the water that goes over the falls; the rest comes from the three upper lakes. But when the upper lakes discharged over Nipissing Pass to the Ottawa valley, Niagara Falls was greatly reduced in volume, having only the discharge of Lake Erie. The great gorge, extending six miles from the falls down to Lewiston, has obviously been made by the cataract itself. According to the laws of erosion it must be true that the falls would cut out the gorge more rapidly when they received the water of all the lakes above, than when they had only that from Lake Erie. We may analyze the course of events best by taking them backwards; thus, Niagara has had the whole discharge of the four lakes above ever since the Champlain uplift shifted the outlet of the upper three. Other things (structure and hardness of the rocks, etc.) being equal, as they are in fact substantially, the gorge made during this time should be wide and deep. If we follow the gorge down from the falls we find that it has this character for a little over two miles. But at this point, just above the railroad bridges, it suddenly grows narrow and shallow.

Returning to the lakes again, we recall the fact that during all the relatively long time of the Nipissing Great Lakes the three upper basins discharged eastward to the Ottawa valley. During that time, therefore, Niagara Falls were small and weak and should have made a comparatively shallow, narrow gorge, and should have cut it out much more slowly. Corresponding precisely with this expectation, we find the relatively narrow, shallow gorge of the Whirlpool Rapids, three-fourths of a mile long, extending from just above the railroad bridges down nearly to the Whirlpool basin. This part of the gorge, then, was made by the river when it had only the discharge of Lake Erie.

The fact that Niagara Falls are slowly cutting their way back through the rocks is well established by observation. Since the first accurate survey by James Hall in 1842, several other surveys have been made, the last in 1890, and it is determined from these that the main or Horseshoe Fall has receded during that period at a mean rate of about 4.17 feet a year. If the two and a fifth miles of wide gorge next below the falls had all been cut at this rate it would have taken about 2700 years to do it. But there is much reason to believe that the rate during the measured period has been considerably more rapid than the average, so that it may have taken two or three times this long to do the cutting. It seems certain that it must have taken less that 10,000 years, but it probably took more than 5,000. There is no basis for an exact calculation, and a more precise statement than this would be no more valuable or reliable. This, then, is a rough measure of the time since the outlet of the upper lakes was shifted from North Bay to Port Huron; since the marine beds of the east were uplifted, and since the sea went out of the Ontario basin.

For the smaller streams to have cut the narrow gorge, three-fourths of a mile long, must have taken a very much longer time, for the weakened cataract would cut back much more slowly. It probably took several times as long to make that part of the gorge as for the longer, wider gorge above. And the time since Niagara began its work must have been still longer, for we have taken no account of the gorge below the Whirlpool basin.

We have seen that the Great Lakes have been very long in the making and that many different forces have been concerned in the work. The giving way of the ice-dams may have been relatively sudden, and perhaps some of the upheavals that affected the lakes were also comparatively sudden; we do not know as to that. It is certain, however, that most of the forces concerned operated only in the slow ways that we see going on now around us. The lake basins are very old; and yet, being newly restored in part by recent glacial obstructions, they are also new. Considering the magnitude of the waters they hold and the great volume of their outlet rivers, the rocky barriers between the basins seem thin and frail. It is only because their youth has lately been renewed by the ice-sheet, with its beneficent contribution of drift, that the Great Lakes of to-day rejoice in the fullness of their strength and proclaim their existence with the voice of Niagara.

FORT WAYNE, IND., April, 1897.

INDEX

NOTE.--In preparing the paragraph on p. 18 concerning the geological age of the surface of Indiana the editor overlooked the statement of Prof. E. T. Cox (Report Geological Survey of Indiana, 1872, p. 138), that upon the hilltops of Pike county he found gravels similar to the *tertiary* deposits of Kentucky and Illinois. Mr. Leverett has examined these deposits and has found them to be quite widely distributed over Southern Indiana as far north as Martin county. He believes them to be of Tertiary or Cretaceous age. This means that the surface of Indiana may be much younger than the geological map of the state would indicate.

Mr. Leverett has recently examined the course of the Collett Glacial River, described on p. 23, and is of the opinion that it did not pass through Clarke and Floyd counties, but emptied into East White River.

Age of Indiana Strata	18, 111	Clear Lake	54
Agriculture	27	Climate of Indiana, Map of.	25
Algonquin, Lake, map of	103	Coal Fields	26, 63
Animals of Indiana	74	Collett Glacial River	22–23, 111
Aspects of Nature,		Committee of Ten, Report of	14
Century of Changes in	72	Cordilleran Ice Sheet	32
Atmosphere	10	Crooked Lake	56
Beaches, Lake	23, 105	Davis, Prof. W. M., Address of	
Beavers of Indiana	74, 75	quoted	7
Birds, Destruction of	81	Definitions of Geography	10-11
Bisons of Indiana	74	Drift, Glacial	96
Black Soil	34	Drift, Materials of	30
Blanchard-Tiffin Moraine	45	Drift, Thickness of	40
Blatchley, W. S., Natural Resources of Indiana	61	Dunes	23
		Eagles of Indiana	75
Boulder Belts	22	Erie Ice Lobe	39, 49
Boulder Clay	21	Erie-Wabash Channel	50, 51–52
Butler, A. W., A Century of Changes in the Aspects of Nature	72	Erie-Wabash Region	42–43
		Drainage of	43, 49
Caves	24	Moraines of	45
Cedar Lake	56	Physical History of	48
Central Plain, The	19	Settlement of	51
Champlain Sea	107–108	Surface and Soil of	51
Champlain Submergence	108	Eskers	22
Channel Lakes	56	Falls and Rapids	23
Cities, The Study of	82, 89	Field Work	13
Clays of Indiana	60	Fishes of Indiana	80

6—8

INDEX

Forests of Indiana 72
Forests, Removal of 78
Fort Wayne 51-52
Fuels 61
Gage Lake 54
Gannett's, H. W., Maps 11
Gas Field 26, 64
Geoanthropology 11
Geobiology 11
Geology, Relation to Geography 12
Geomorphology 11-13
Geophysiology 11
Geography, College Courses in . 15
Geography, Divisions of 11
Geography, The New 9
 Summary of 15
Geographical Congress, Third International, Resolutions of . . 10
Gilbert, G. K., His Work in Erie-Wabash Region 44
Glacial Boundary 26, 28, 33
Glacial Gathering Grounds . . . 31
Glacial Lakes, Map of 98-99
Glacial Period 95
Glacial Succession 32
Glaciated Rock Surfaces 31
Gorges 23
Great Lakes, A Short History of . 90
 Erosion in Region of 93
 Origin of 92
Greenland Ice Sheet 32
Hettner, Prof., View of Geography 10
Hydrosphere 10
Ice-Dams 101
Ice Invasion, First 32
Ice-Lobes 96
Ice-sheets 95, 102
Improvement of Geographical Teaching. 7
Indiana: A Century of Changes in the Aspects of Nature . . . 72
Indiana, Animals of 74
 Building Stones of 67
 Clays of 69
 Climate of 24
 Coal of 63
 Drainage of 20
 Elevation of 17
 Forests of 72

Glacial Deposits of 28-29
Geological Structure of . . . 18
Hills of 21
Indian Villages of 74
Iron Ores of 70
Lakes of 23, 53
Limestones of 67
Moraines of . . 19, 22, 38, 45, 46, 96
Natural Gas of 64
Petroleum of 65
Physical History of 18
Physiographic Features of . 21
Physiographic Regions of . 18
Plains of 21
Population of 26, 27
Position and Boundary of . 17
Pre-Glacial Surface of . . . 94
Rainfall of 24-25
Resources of 26, 61
Sandstones of 68
Settlement of 27
Soil of 66
Temperature of 24, 25
Topographic Map of . . Frontispiece
Vegetation of 25
Indian Villages of Indiana . . . 74
Interglacial Interval, First . . 34, 36
Iron Ores of Indiana 70
Iroquois, Lake, Map of 103
James Lake 57
Kankakee, Basin of 19, 23
Kankakee River 21
Kankakee Glacial Lake 39
Kames 22
Keewatin Ice-sheet 32
Kettle-hole Lakes 54
Knobs 22
Labrador Ice-sheet 32
Lagrange County, Lakes of . . 55
Lake Algonquin 103-104
Lake Chicago 98, 101
Lake Duluth 98
Lake Erie 90-91
Lake Huron 90-91
Lake Iroquois 103-104
Lake Maumee 50, 98, 100
Lake Michigan 90-91
Lake Ontario 90-91
Lake Saginaw 99, 101
Lake Superior 90-91

INDEX 113

Lake Warren	101, 105	Pigeons of Indiana	75, 77
Lake Whittlesey	99, 101	Resources of Indiana, Natural	26, 61
Lakes, Glacial, Classification of	53	Ritter's Idea of Geography	9
Lakes, Ice-dammed	97	Saginaw Ice-Lobe	49
Lakes, Life History of	59	Salamonie-Blue Moraine	46
Lakes of Indiana, Morainic	53	Sandstones of Indiana	68
Lapparent, Prof.de, View of Geography	10	Shades of Death	23
		Shriner's Lake	56
Leverett, F. Glacial Deposits of Indiana	29	Silts, Glacial	36
		Silver Hills	22
Limestones of Indiana	67	Sinkholes	24
Lithosphere	10	Soils of Indiana	27, 66
Long Lake	56	St. Joseph River	44
Loess	21, 35-36	St. Mary's River	44
Mackinder, H. J. View of Geography		St. Mary's-St. Joseph Moraine	45
phy	10	Stone Quarries	26
Marshes	24	Stones, Building	67
Maumee Lake	20, 23	Striae, Glacial	28, 29, 31, 33, 40
Maumee River	43-44	Taylor, F. B. on the Great Lakes	90
Michigan, Lake, Basin of	19	Terraces	23
Mississinewa-Eel Moraine	46	Terre Haute, a Study of	83
Moraines	19, 22, 38, 45, 46, 96	Turkey Lake	24, 58
Moraine Topography	47	Turkeys of Indiana	75
Neolithic Man	76	Valleys, Buried	94
Neumann, Prof. View of Geography		Valleys of Indiana	23
phy	10	Wabash River	20
Niagara, Falls of	100	Wabash River, Scanty Knowledge	
Niagara River	101	of	8
Nipissing Beach	106	Wabash-Aboit Moraine	46
Nipissing Great Lakes	105, 107	Walden Pond	55
North America, Glacial Map of	33	Warren County, Glacial Deposits	
Northern Plain, The	19	of	40
North West Territory, Seal of	76	Weed Patch Hill	22
Ohio Valley	19	White Occupation, Results of	75
Ohio Slope, The	19	White River	20
Paroquets of Indiana	75	Whitewater River	20
Penck, Classification of Morainic		Williamsport, Glacial Striae near	40
Lakes	54	Wisconsin Boundary	33, 36-37
Petroleum Field	26, 65	Wisconsin Ice Invasion	37
Physiography	12	Wyandotte Cave	24

www.ingramcontent.com/pod-product-compliance
Lightning Source LLC
Chambersburg PA
CBHW020139170426
43199CB00010B/809